Chatham

The Life of a House

By Ralph Happel

Philadelphia, Pennsylvania
Published for Fredericksburg and Spotsylvania National Military Park

By Ralph Happel
Design & Illustrations by Frank Fretz

ISBN #0-915992-30-2
Library of Congress Catalog Card Number 83-82829

Published by Eastern National Park & Monument Association. Eastern National promotes and aids the historical, scientific and educational activities of the National Park Service. As a nonprofit Cooperating Association authorized by Congress, it makes interpretive and educational material available to park visitors by sale or free distribution. It also supports research, interpretation and conservation programs of the Service.

Contents

A Quiet Mount

BY APRIL OF 1863 men on horseback had been brandishing swords, at intervals, for two years. Though the countryside was losing, nobody seemed to be winning this American Civil War. Looking forward to dry battle routes after the wet torpor of winter quarters, generals were awakening to the peeps of the herring frogs and reaching for maps.

The calls of spring transcend generals. April is the time when many folk long to go on pilgrimages, preferably peaceful. Thus on April 6, seizing the day and waving neither sword nor tract, a couple of Yankee chaplains came riding down on ambling nags from their camps in Stafford County, Virginia, to the left bank of the Rappahannock, the river dividing their army from the Rebels.

One of the riders contributed dispatches to a Congregationalist paper in Massachusetts; he was seeking copy as well as having a day off. After the obligatory denunciation of Virginia's roads, still not dry, in a subsequent article he told his home-front readers that he found himself that April Monday on the heights opposite the town of Fredericksburg where the Federals had failed in December of 1862. His regiment had not been at that battle. Across the narrow river shining in the sun Confederate sentries could be seen walking their beats. The men of these opposing armies usually refrained from shooting at each other between battles. The spires of the battered town yet stood, as did a brick mansion behind him.

An officer there told the visitors about this house. It was called the Lacy House, after its owner, absent in the Rebel Army. The place had served as headquarters, artillery position, signal station, staging area, and hospital. The stark building, obviously once an abode of comfort, looked now over bleak grounds. Shrubbery and most of the trees had disappeared. This was a wood-burning war. Flower beds, lawns and paths were trampled mud.

Inside the war-torn house, the chaplains saw a lonely relic of peacetime, an object to haunt a family man. What infant sportsman, or commander perhaps, had led thereon the hunt or charge? No ghost itself, prosaic, solid, a child's rocking horse stood quiet by a doorway to desolation.

"That elegant and highly improved seat"
1. The Fitzhughs: Colony & Revolution

THE RAVAGED HEIGHTS so deprived of trees in the 1860's once stood deep in the New-World forest, just below the first falls of the Rappahannock. In the summer of 1608, exploring the rivers of Virginia, Captain John Smith, the advance man of English settlement, cast a warm eye on that greenery, but more than a hundred years would pass before the town of Fredericksburg would rise on the right bank below the falls and Falmouth on the left at the falls, and those English outposts would grow for almost half a century before the construction of the house the chaplains visited.

A Lacy did not build that house. Another Virginian of Anglo-Norman descent, William Fitzhugh, built it at the eve of the American Revolution and called it Chatham in honor of William Pitt, Earl of Chatham, whom Virginians esteemed.

Settlement had pushed up the neighboring Potomac earlier than along the Rappahannock. The first Virginia Fitzhugh, also a William, seated himself on the Potomac on land now in King George County (then in Stafford) during the 1670's. Springing from a line of substantial burghers, he set out to be an aristocrat in the new country. Proud of his coat of arms (many families in trade were so entitled), he prospered and used silver forks to eat his meat when other settlers were still spooning their broth with pewter, and he ordered a coach for use on primitive horse roads. Planter, politician, land acquirer, and attorney, that William Fitzhugh bequeathed a heritage of power and land to a vigorous progeny. Reaching over to the Rappahannock, he acquired the acreage of future Chatham in the 1690's. Among previous patentees had been two Claiborne brothers; their legacy is the name of a little stream running into the river not quite a half mile below Chatham house. William the immigrant's 700 acres here descended to the fortunate great-grandson and namesake. Early Fitzhughs built a gristmill on Claiborne's Run and operated a plantation on the property, but a significant house did not rise until William Fitzhugh of Chatham built that dwelling.

Born at Eagle's Nest on the Potomac in 1741, the inheritor of large properties in several counties, this William Fitzhugh, the wealthiest of the extended family, married Ann Randolph (whom he called Nancy) of Henrico County in 1763, she not quite 16 and he 21 years of age. They were residing at Somerset plantation, adjoining Eagle's Nest, when they decided to seat the Rappahannock acreage. Fredericksburg was opening up the back country while retaining the flavor of Tidewater life. The neighborhood of the little business and social center beckoned. Wealthy in land, William Fitzhugh sold 10,000 acres to acquire the cash for building Chatham.

Virginia's clay and forests provided bricks and timber. Skilled builders translated the Georgian style of England into a simpler, scaled-down elegance with the aid of capable workmen, some of whom occasionally felt they were more than worthy of their hire. In the early stages of Chatham's construction, 1769, a slave called William Hanover, "by trade a good house carpenter and joiner," put on his "kersey surtout coat, with flat metal buttons," packed up the rest of his extensive wardrobe and took off down the river. Neither this defection, however, nor the great flood of May 1771 that caused much damage in Fredericksburg and Falmouth stopped the completion of Chatham. By the autumn of 1771 William and Nancy had taken occupancy. In November a child, Lucy Fitzhugh, was born to the couple.

As they settled in, they furnished a dwelling imposingly long, though not wide, consisting of a two-story center portion and one-story wings. The brickwork was Flemish bond, each course made up of alternating lengthwise bricks and headers, with headers glazed. Just off the northern end stood a detached brick gable-roofed laundry, and off the southern end a similar kitchen. The usual offices and other buildings gave the typical plantation appearance of a hamlet. Indeed, with overseer's house and slave quarters, such places truly were hamlets.

A capable agriculturist and horse breeder, William Fitzhugh had received from private teachers the early education of a planter's son, but did not, as did many of his peers, attend the College of William and Mary. His father, Colonel Henry Fitzhugh, had exceeded colonial higher education by matriculating, aged 15, at Christ Church College, Oxford, in 1722. (The precocious Henry died in his middle thirties when William was an infant. The widow Lucy Carter Fitzhugh married Colonel Nathaniel Harrison and continued at Eagle's Nest.)

William Fitzhugh of Chatham

Transplanted planter William Fitzhugh, who already knew the merchants and gentry of the Fredericksburg area, gained a valued neighbor to the south, James Hunter of Pine Grove, proprietor of an iron works at Falmouth. The owner of Ferry Farm, next to Hunter, was of greater note, but Colonel George Washington no longer lived at that farm, fields of which he rented to Hunter and Fitzhugh. About 1775, Washington, whose mother had moved from Ferry Farm to Fredericksburg, sold the property to Hugh Mercer, a Fredericksburg physician of Scottish birth, a survivor of Culloden, and Washington's friend since their service in the French and Indian War. Mercer became Fitzhugh's doctor and apothecary in 1771.

William Fitzhugh, sociable and intelligent, proved apt politically. The autumn of his removal to Chatham saw him elected a member of the Virginia House of Burgesses from King George, Chatham then being in that jurisdiction, which was then a narrow county along the north shore of the Rappahannock. Washington, a fellow member, was a leader in opposition to British policies. The

Chatham, Front Facing River

1770's, after the turbulent '60's, voiced Virginia's increasing resentment. The Chatham planter took horse to Williamsburg for sessions of an uneasy House.

Domestically, the year 1773 fretted and distressed William Fitzhugh. Mrs. Fitzhugh, carrying another child, suffered a troublesome pregnancy. In October, a month shy of delivery, she was still confined to her room after upward of five weeks of a fever and cough. Fitzhugh left the plantation that fall only once, to go over to the Potomac to see his mother, lying dangerously ill; she died before the year was out. Nancy bore Betty Randolph Fitzhugh late in November.

On the public scene, all through the colonies the colonials winced under pressure. In Massachusetts anger found action in the grim festivities of the Boston Tea Party late in 1773. The home authorities, their dignity stung and patience torn, retaliated with severe measures; one, the closing of the port of Boston, caused privation to the populace. There and everywhere, people realized in early 1774 that affairs had come to a matter of submit or fight. Opposition to the tax on tea became a rallying point.

The ladies of Fredericksburg vowed to take no tea. There were other beverages, however, and other concerns, George III or no George III. Everybody locally was wondering how William Fitzhugh's gray mare Kitty Fisher would run at the May Fair races.

She had won a sweepstake of 50 guineas at the last Annapolis meet. The gray mare seemed not what she used to be, in the event, and lost to a bay mare named Miss Alsop. A few days later, horse players heard a poignant text at a special service on June 1, a day set apart by the House of Burgesses as a time of fasting, humiliation and prayer, and the imploring of divine interposition for averting assaults on the rights of America. The Reverend James Marye rang the spaces of St. George's with *Psalms,* XII, 1, "Help, Lord; for the godly man ceaseth; for the faithful fail . . . ." Some of the gamblers could not distinguish the sound of *man* from *mare.*

In Fredericksburg's October races, form picked up for William Fitzhugh. Kitty Fisher won the Town Purse of 50 pounds. His stallion Regulus won the Jockey Club Plate, 100 guineas, and his chestnut gelding Volunteer took the Town and Country Purse, 50 pounds. The Fredericksburg Jockey Club presented Fitzhugh with a pair of silver tankards. Besides racing, citizens and visitors during that week watched puppet shows, rope dancing and tumbling. Over the river, the gentry jubilated over Fitzhugh's wine.

Then sorrow visited Chatham; on October 10, William and Nancy Fitzhugh mourned the loss of the infant Betty. "In the midst of life . . . ."

The Americans were also in the midst of the death of a colony, but ordinary affairs, less vociferous than politics and less exciting than an occasional horse race, went on. Merchants bought and sold; workmen labored at their appointed tasks. At Chatham, Ann Fitzhugh supervised housekeeping, the welfare of family and servants, and her developing flower gardens; William and his overseer attended to planting and harvesting tobacco, wheat, barley, Indian corn, and clover. Orchards were planted. Rows and beds of small fruit for preserves, shortcake and tarts grew to maturity. On the river the men at the fisheries netted the spring fish for the salting, with the occasional sturgeon to give a planter's table better roe even than that of herring and shad. At the mill on Claiborne's Run the water turned the wheel for grinding flour and the meal of Indian corn, a New-World staple; while the mill of dissent between Mother Country and Colony ground out bitterness.

In March of 1775 William Fitzhugh attended the Second Virginia Convention at Richmond and heard Patrick Henry's rousing (to the Whigs) and demagogic (to the Tories) give-me-liberty-or-give-me-death speech. That year the Second Continental Congress, meeting in Philadelphia, made George Washington commander of the American forces, as the Revolution began. The Chatham squire in his lesser range became one of the commissioners for the establishment of a gun factory at Fredericksburg. For Fitzhugh war did not come to, in the phrase of a later century, guns or butter. He could, in March of the fated year 1776, write to his uncle Landon Carter of Sabine Hall on the lower Rappahannock that he hoped that the troublesome fish bone was out of Carter's throat, going on to announce the delivery of ten fine pigs from "Little Peg," who managed to kill three of her litter; the Colonel should have some of these noted Lilliputian hogs; Fitzhugh hoped to visit Sabine Hall soon; nowhere else did he feel better, because of the good claret or the apple brandy toddy or the "Nassaws Mixture."

Later in the year the taste of good living soured. On December 20, writing again to his uncle, this time from Williamsburg, by candlelight beside a poor fire, "and with a very weak Eye"—he had but one, having lost the other in a childhood accident—he relayed sad news from Philadelphia: "The Enemy within a few Miles of that City, and our worthy General flying before them, with a few fatigued and shatter'd Troops. Now is the glorious Opportunity for Virginians to stand forth or we are undone, for Virginians [a]lone must save America . . . I would with Pleasure part with my last Shilling to preserve my Liberty."

A few days later, the worthy General stood forth—in fact, rowed forth—crossing the Delaware on Christmas night with Virginians and some other colonial troops to surprise the Hessians at Trenton. He then defeated another British force, January 3, 1777, at Princeton. Washington's victories changed both the immediate situation and the course of the war. Hope and faith returned to the hearts of the Whigs, who missed, however, a valiant soldier; Dr. (now General) Hugh Mercer fell mortally wounded at Princeton. Fitzhugh lost a neighbor who never had the opportunity to settle down and enjoy Ferry Farm. Congress voted to educate Mercer's son Hugh, "the child of the Republic," and Fitzhugh would be one of the trustees.

Fitzhugh, far from his last shilling, had during the year 1776 advanced money for payment and subsistence of militia and the delivery of muskets and bayonets from James Hunter's Iron Works. In 1777, under an act for the encouragement of iron works, he was appointed a commissioner to foster the expansion of Hunter's works, the Rappahannock Forge, to furnish products for the army and navy. Both the Fredericksburg gun factory and Hunter's contributed materially to the Revoluntionary cause. That year also Fitzhugh became a Burgess from Stafford County.

Chatham was now in Stafford. The boundaries had been changed; Stafford and King George now ran across Virginia's Northern Neck from Rappahannock to Potomac, and the eastern part of old Stafford (including William Fitzhugh's Potomac plantations) had become part of King George.

In that year of change, when British ships came up the Potomac, he and relatives lost slaves by desertion to the enemy. A master's loss may have been the slave's gain, but the house servants of Chatham joined master and mistress in mourning the earthly loss of Lucy, September 29, 1777.

In 1779, a delegate to the Continental Congress, William Fitzhugh journeyed to Philadelphia. The weariness of office, the press of business, going to races, as to Petersburg in the spring of that year, never precluded Chatham civilities when the Fitzhughs were in residence. Before attending those races, Fitzhugh entertained General George Weedon, ex-host of Weedon's Tavern across the river (where Fitzhugh enjoyed billiards), with a saddle of mutton, Madeira, claret, and whist. Though a true Whig, the Virginia gentleman Fitzhugh could also extend courtesies to an enemy officer. British Major General William Phillips, whom Thomas

Jefferson regarded as the proudest man of the proudest country in the world, humbled himself enough to be most grateful to Mr. and Mrs. Fitzhugh for kindness shown when he visited Fredericksburg as a prisoner of war. He tried to reciprocate, after release, by gaining the parole of two of their relatives. Untoward events, however, interposed.

When Phillips and Benedict Arnold forayed in Virginia in 1781, Fredericksburg authorities convinced Governor Thomas Jefferson of the danger posed to Hunter's Iron Works and the Gunnery, especially since Phillips had had the opportunity of seeing the importance of the Fredericksburg area. British vessels on the Potomac, a short distance overland, frightened the townspeople, but no attack came. The British preferred Southside Virginia, where General Phillips died of fever at Petersburg on May 13, thus escaping the future indignity of Yorktown.

2. *The Fitzhughs: In the Early Republic*

WITH THE COMING OF PEACE and the confusion attendant upon the birth of a nation and the change of colonies to states, William Fitzhugh eased back into the life of a country gentleman, never having fully left it. He did not entirely eschew politics, serving in the state legislature in the 1780's; Richmond, the state capital, was not too far from home and family. The Lord's business also occupied him; the good churchman in 1785 and '86 attended conventions looking to the reorganization of the Episcopal Church after the Revolutionary disestablishment. Though a son was still denied him, he welcomed with Nancy the arrival in 1783 of Ann Randolph Fitzhugh, little Nancy.

Horses again received their just attention. In November of 1785 Fitzhugh wrote to his friend George Washington that the horse Tarquin had recovered laurels lost in Alexandria. Presumably the General knew of those lost laurels. Fitzhugh sold that horse five years later to his wife's cousin, Thomas Jefferson. Meantime, the Chatham Fitzhughs and the Mount Vernon Washingtons remained on intimate footing. "I had rather," Fitzhugh once stated, "be an attendant on that great man than the first Potentate in Europe." Fitzhugh visited Washington, and Washington visited Fitzhugh. They exchanged seeds and plants and corresponded. Fitzhugh offered several does for a deer park the General contemplated, two from Eagle's Nest and one from Chatham. The Chatham doe, perfectly tame, was continually harassed by a pack of neighboring hounds and had to take shelter in the kitchen.*

Trespassers, canine and human, habitually violated the calm of Chatham, forcing Fitzhugh to forewarn all persons from hunting and shooting on his land. This notice did not prevent several years later the theft of two work horses just when plowing was about to begin.

*Presumably the Chatham doe's descendants were among the herd browsing Washington's shrubbery when his park fences became derelict during his presidency.

His family grew with the birth of Martha Carter (Patsy) and Mary Lee (Molly) Fitzhugh in 1786 and 1788. In a few years Patsy began work on a sampler which gives the birth and marriage dates of Mr. and Mrs. Fitzhugh and recites the births and deaths of Lucy and Betty. Having finished with that information, Patsy threaded a verse to the dead infants:

Here Innocence and Beauty lie, whose Breath
Was snatch'd by early, not untimely Death.
Hence

at which untimely point she stopped, leaving the work unfinished. She died, September 29, 1793, aged 7.

A sister, probably Molly, took up the fallen needle to finish the verse:

did they go, just as they did begin
Sorrow to know, before they knew to sin.
Death, that does Sin and Sorrow thus prevent,
Is the next Blessing to a Life well Spent.

The sister, stitching dutifully in the house on the hill, recorded the others, explained about Patsy, gave Patsy a verse, and at the bottom noted:

William Henry Fitzhugh born March 9th 1792

A male heir had arrived at last.

In 1793, Fitzhugh bought 588 acres to the rear of the original 700 Chatham acres. His affinity for horses did not diminish; such stallions as the True Whig and Chatham added to the good lines of Virginia. Nor had Fitzhugh's reputation as an expert on the points of a horse been forgotten. In August of 1798 Washington asked his friend to look out for a good riding horse for him; he would especially need it if called to the field. (The young republic was having difficulties with Revolutionary France; Washington was to command the army, but nobody expected the old man himself to take the field. At any rate, the French never invaded.) As to color, the General preferred white. Fitzhugh found a horse perhaps suitable and respectfully suggested that a horse the General had purchased was a little long in the tooth; its color was not mentioned.

An agriculturist grows weary in his age. In his fifties, well-advanced years for that time, William Fitzhugh began to find burdensome the operation of Chatham and his vast holdings of Ravensworth in Fairfax County. Crops failed, as they will, by too little or too much rain, or rain on one field and not another. International situations, as before and since, affected markets; peace between France and Austria lowered the price of grain. Somerset and Eagle's Nest he had given over, but not deeded, to his nephew Captain Benjamin Grymes, a veteran of the élite corps of Washington's Life Guards. Like his uncle, Grymes revered George Washington. On occasion in the years following the Revolution a sloop would tack some fifty miles up the Potomac to Mount Vernon. The sail would clatter down on its rings as the anchor dropped; a small cannon would fire, and a messenger would row ashore in the sloop's batteau to deliver a large rockfish or other seafood from Eagle's Nest. Less dramatically, summer after summer, Ben Grymes kept Chatham supplied with fish and crabs from the saltwater Potomac. Fitzhugh was grateful, but Eagle's Nest and Somerset still required his attention; Ben Grymes himself was a burden; he had yielded to alcoholism. Delirious and violent when drinking, he killed a Fredericksburg merchant and was confined a while at the Williamsburg Insane Asylum during the mid-nineties.

There was often sickness at Eagle's Nest, for which Fitzhugh paid. (Once in trying to get a doctor sent there, Fitzhugh noted "they will not go a Distance.") Chatham too had its illnesses: agues, fevers, severe colds, and pains of all sorts. Winter induced rheumatism, and complaints of the bowels came with the summer, as did also malaria, which the Tidewater people fled by going to the upland springs. Smallpox was not seasonal. One year several slaves at Ravensworth died of that old malady.

Moreover, Chatham had become a sort of free tavern for the gentry of the lower counties going north and returning. Sometimes there was not a spare bed in the house. Randolphs, Lees and other relatives and friends abounded. Debts plagued the host worse than sickness; he called himself a slave and worse than a slave, laboring with body and troubled in mind.

Though he shuddered at the idea, he determined to sell Chatham. The sale would more than settle his debts. He decided to buy a house in Alexandria, only ten miles from Ravensworth. That town dwelling and the 12,000 acres of Ravensworth he considered a sufficient legacy for his boy; his daughters and wife could easily be provided for. In late 1796, he bought the Alexandria house, a fine brick structure with outbuildings and a pump of excellent water.

In his effort to simplify and consolidate his life and affairs, Fitzhugh planned to sell not only Chatham, but other lands in Stafford and elsewhere. Chatham, he stated in an advertisement of early 1797, was a brick house delightfully situated and containing nine commodious rooms, exclusive of a spacious hall or entry, two pairs of stairs, convenient passages, and good dry cellars.

The grounds comprised pleasure and kitchen gardens, interspersed with scarce trees, choice flowers and flowering shrubs, fruit trees and berries. Adjacent were orchards of peaches, apples and pears. Features of the grounds, said Fitzhugh, were turfed slopes created by great labor and expense.

Detached from the mansion house were offices of all kinds: kitchen and larder, housekeeper's rooms and laundry, storehouse and smokehouse, all of brick, dairy and springhouse of stone, also stables for thirty horses, coach houses for four carriages, a large farmyard with barns and granary, cow house with stalls for thirty-six grown cattle, sheds for fattening veals, muttons and lambs, and sheds for sheep, also an overseer's house, blacksmith shop, and quarters to accommodate more than fifty laborers.

Fields of red clover and orchard grass provided ample hay. Arable land was laid out for crop rotation.

Washington in 1775.

Steel Engraving (1857) from
a Depiction by Alonzo Chappel

The mill on the premises, advantageously rented out, furnished with modern machinery, was built of freestone, as was the miller's house. There were quarries of freestone near the river, a well-paying ferry from Chatham land to Fredericksburg, and two or more valuable fisheries.

The roads were good, neighborhood sociable and genteel.

Genteel company, making use of the good roads, kept coming to Chatham, as Fitzhugh, filled with melancholy, prepared to leave. The family began packing up in October of 1797. Mrs. Fitzhugh, bothered by a leg affliction and devoted to Chatham, disliked the move even more than her husband did. Fitzhugh had a bad cold. They hung on a while. By the middle of December, they were in residence in Alexandria.

During succeeding years the family would see more of the Washingtons. In the middle of November 1799, the General went to church in Alexandria and dined with the Fitzhughs. That was his last foray from Mount Vernon.

However much furniture the Fitzhughs removed from Chatham, they evidently left some or brought some back, because for years after they moved they (sometimes only William) continued to return to the Rappahannock to stay a while, Chatham remaining unsold. Nancy failed to sign a deed of release on other Stafford property and may have hindered the disposition of Chatham. Company still came; once in the summer of 1799 the Fitzhughs gave breakfast, dinner and lodging to twenty-three, and the day before dined upwards of forty. In that summer, visitors consumed during the first half of August twenty-one calves, sheep, lambs, and shoats from the place, besides a number purchased, fowls beyond counting, and three sturgeons. All the while William Fitzhugh was shy of currency.

In October 1800, young Nancy was married at Chatham to Judge William Craik, son of Washington's physician, Dr. James Craik. Briefly, life and joy visited the old homestead.

On August 10, 1805, Ann Fitzhugh died, at fifty-eight. On October 18, William again advertised the sale of Chatham, "that elegant and highly improved seat." On May 9, 1806, it was purchased by Major Churchill Jones.

Ann Craik died, without issue, in October of 1806, and William, sixty-eight years old, followed this daughter, December 19, 1809. An obituary suggested that the prayers of the many warmed by his fire, fed by his bounty and relieved by his charity would speak for him at the tribunal of eternal happiness.

He left, among various bequests, Somerset and Eagle's Nest to his three grandnephews, the sons of Captain Grymes, already deceased. William Henry Fitzhugh presumably would carry on at Ravensworth and Alexandria and would marry and continue William's male line. He married, but died young without children.

Mary Lee (Molly) Fitzhugh had married, 1804, George Washington Parke Custis, of Arlington, Martha Washington's grandson and the General's adopted son. Molly's daughter, Mary Anna Randolph Custis, would grow up to marry a boy almost three years old at the time of William Fitzhugh's death, the son of his cousin "Lighthorse Harry" Lee—Robert Edward Lee. R.E. Lee would live during his boyhood at William Fitzhugh's Alexandria house. This building, 607 Oronoco Street, is now (1982) a house museum.

Chatham, Laundry House

"All betokened the Old Dominion"
1. The Jones Brothers & Martha

BREVET MAJOR CHURCHILL JONES, purchaser of Chatham house and 1288 acres, former cavalry captain in the Revolution, got along well with his brother William, a Tory. The Revolution had not been for them a war between brothers. After the war William and the new order accepted each other; his knee breeches—he scorned the trousers of the republic—became a mark of gentlemanly peculiarity. He and Churchill were neighbors in the area called the Wilderness about fifteen miles west of Fredericksburg, he at Ellwood plantation and Churchill at Woodville. They had come to the Piedmont from Virginia's Chesapeake shores just before the Revolution.

Churchill Jones' first wife died early. Then he married a widow who died in 1802. In June of 1805 he married Mrs. Martha Selden Douglas, wealthy relict of a Petersburg merchant. The couple moved to Chatham after the 1806 purchase, though spending some time each year at Woodville in Orange County.

Martha Jones, a lady who knew her own mind, strove to bring Chatham back to its old splendor, indoors and out. She set her workmen to beautification of the grounds. She kept a well-ordered house with cupboards under lock and key. The kitchen came in for particular surveillance. Her file of receipts told, among many things, how to preserve strawberries whole, to make red currant jelly, the best shortbread, buns, gingerbread, catchup, to preserve squash, to pickle beef. She also knew how to make ink and stop nosebleeds.

She did not know how to curb her temper or accept life (at least without a fight) on any terms but her own.

She and the Major had no children by their union or previous marriages, but two nieces under the Chatham roof were treated as daughters: Ann Elizabeth Selden and Martha Ann Elizabeth Macmurdo. Both children answered to the name Elizabeth.

Churchill Jones

Following the death of Elizabeth Macmurdo's mother, her father had remarried and produced a large family for whom Aunt Martha did not care; so Aunt Martha had snatched up Elizabeth Macmurdo to rear. Though pridefully cherishing both her Elizabeths, she felt the Macmurdo child to be her very own.

In 1812, Elizabeth Macmurdo, aged 20, decided to visit her father and stepmother in Richmond. There she met Patrick Gibson, a rich merchant, much older than she and a widower with children. Mr. Gibson soon appeared at Chatham to ask permission to court Elizabeth, her father having consented. The suitor was summarily rebuffed. Martha, a successful matcher of other kin in the past, refused to believe that Elizabeth could be deluded enough to consider taking on such potential trouble. Elizabeth, confined to Chatham, remained demure and lay low, having promised under duress to marry nobody without Martha's approbation.

At the suitor's request, Elizabeth's father travelled up to Chatham. He faced down the aunt, took Elizabeth to Richmond and then to a relative west of that city. Martha Jones began pulling strings all over the place, obviously failing in her manipulations. She swore never to see Elizabeth again if the girl married Gibson. The girl, mindful of the aunt's devotion and of a happy and privileged childhood, wrote a sad letter to Major Jones, who wrote a sad letter back. He was suffering from an affection of the head caused by his wife's rage, a wrath now lapsing into melancholy.

In early 1813, Mr. Gibson journeyed to Fredericksburg and sent a note to Chatham requesting an interview. Mrs. Jones wrote back denying the audience and stating that not "for Ten Thousand Worlds" would she break her vow of never speaking to Elizabeth again if Elizabeth married him. She begged Mr. Gibson not to probe further the wound in her afflicted bosom.

On the 27th of February, 1813, Patrick Gibson and Elizabeth Macmurdo were married.

A cousin of Elizabeth's visited Chatham in April and then reported to Elizabeth that Aunt Jones wept and wished all happiness to the couple, admitting that Mr. Gibson had always acted properly. Major Jones now seemed to be the obstacle to reconciliation. In any event, ten thousand worlds to the contrary, the families became reconciled by autumn. Elizabeth thereafter was a frequent visitor to Chatham and a loving correspondent between visits.

This tempest in flowered and cozied teapots made no great noise beyond the Chatham and related parlors, but the second war with Britain shook everybody in the Fredericksburg neighborhood in 1814. The burghers and their rural friends had feared invasion during the Revolution. Now once again they all but smelled the Redcoats' gunpowder when British ships sailed up the nearby Potomac.

On August 25, 1814, writing to her "dearest Child" Elizabeth Gibson, Martha Jones said she feared the city of Washington had been in flames the night before when the Chatham family heard cannon and saw a glow in the northern sky. British ships were lying off Potomac Creek in sight of Salvington, her relative Cary Selden's place overlooking the Creek, only seven miles from Chatham. The enemy might march over any time. She and the Major had sent off (presumably to Woodville) five wagon loads of possessions, were still packing and were ready to flee at the first alarm, leaving Chatham to look after itself, possibly to be burnt. Meanwhile, she asked Elizabeth to keep, for the time being, the bonnet bought for her, which she hoped was not like the one sent down; that one was too youthful and did not have a wide enough brim to shade the face.

Mrs. Jones was right about Washington. A British force had marched over to the new national capital from the upper Chesapeake Bay and had fired public buildings on the 24th. The ships on the Potomac, not going above Alexandria, had levied tribute on that town and gone downriver without disembarking anybody to march on Fredericksburg. Aunt Jones' epistle of fear ended in bonnets, and the second putative threat to Fredericksburg ended in the return of Chatham's goods and of the local militia to daily tasks. Martha could now return to family affairs.

She approved and furthered a match between the other Elizabeth, Miss Selden, and John S. Knox, a Fredericksburg merchant. They were married at Chatham in October of 1816.

Worse, locally, than any effect of war with England came a black cloud in May of 1818, forming slowly for the benefit of any prescient enough to close outside shutters, to hurl a devastating weight of hail on Fredericksburg and Falmouth. Major and Mrs. Jones and the William Jones family, then visiting Chatham, were out at the time of the storm. On returning, they found remaining unbroken only forty-seven panes of glass (some cracked) out of five hundred. Rooms and furnishings were wet and battered. Outside, the gardens and crops were cut down, shrubbery and trees stripped of leaves—no fruit, no oats, no wheat, no beauty! Major Jones calculated that their loss would reach 2000 dollars. He bore all as a Christian, never fretting at providence. His wife bewailed her inability to be like him, "but alas, I am Martha still," she cried.

Chatham duly recovered and during the next several years Martha's interest in the lives of others cheered her up. With no more young Chatham relatives to oversee, she turned her attention to Ellwood and the William Jones family.

William Jones had one daughter, Hannah, whose husband David Williamson, a Scottish merchant of Fredericksburg, died in 1818, leaving a daughter Janet, defective in mind, hearing and sight. Then along came Judge John Coalter, a mature widower with chilren, to pay court to the widow. Aunt Martha considered this a good development all around. She liked Judge Coalter, a poor boy of Augusta County in the egalitarian west, who had risen through probity and merit. He had been a protégé in Williamsburg of St. George Tucker, a noted teacher of law, before marrying and going to Staunton to practice law himself. His wife died in childbirth. So did a second. Later at the Sweet Springs of Monroe County (presently in West Virginia) he found Frances Tucker, whom he had taught as a young tutor in her father's house. They were married in 1802. In 1813, after a happy marriage producing several children, she died at the same springs.

In his middle years, a judge of the Virginia Court of Appeals, Coalter pressed his case judiciously, abetted by Martha Jones. He married Hannah Jones Williamson in February of 1822 at Ellwood. Martha attended the wedding, leaving the Major, who did not feel well, at Chatham.

The Major felt his years. Nevertheless he pushed work on a toll bridge he was building across the Rappahannock to the foot of Fredericksburg's William Street. That bridge may have caused his death. He died in September of 1822, aged 74, at his summer home in Orange County, of a bilious fever, probably typhoid, brought on, some said, by exposure in supervising the bridge construction. (This bridge went with a flood of 1826. The rebuilt bridge was sold in 1847 with reservation of free passage for Chatham family and servants.)

Churchill Jones left about seventy slaves at Chatham and a good bit of silver, glass and furniture. A sale of perishable property in December bore out the immemorial dottiness of auction goers. Six pickle leaves (bay leaves?) fetched 15¢ for the lot, and a blind horse brought $6.00. He freed a number of slaves by his will and left his real property to his wife for life; after her death it would go to his brother or his brother's heirs. Martha was to be executrix and William executor. Martha, dissatisfied, turned against William

Mrs. Hannah Coalter

Jones and family, the relatives she had so often visited and who had so often come to Chatham. She unburdened herself to her now respected in-law Patrick Gibson, erstwhile prober of her afflicted bosom, confiding that she had not realized what an ignorant and obstinate man William Jones was. She could take a third of all property—anyway, Chatham was too big for her to manage alone.

Finally, on December 28, she entered into an agreement with William Jones, giving up all rights for an annuity of $800 for life. Whereupon she moved to Richmond and bought a house there. As the world turned, she again grew fond of William Jones and kin and spent visits at Ellwood and Chatham.

Patrick Gibson died in 1827. Elizabeth Gibson and Martha leaned on each other until Martha's death in 1838. She was buried from Elizabeth's house. Elizabeth was to live until 1867.

2. The Coalter Connection & Betty Lacy

WILLIAM JONES transferred Chatham to his son-in-law John Coalter late in 1825, the deed being executed in 1829. Coalter agreed to keep up the annuity to Martha Jones.

Judge and Mrs. Coalter and his grown children soon became a part of the social life of the Fredericksburg neighborhood. They were especially friendly with the Mercers and the Mercers' Patton relatives (whose descendants would include a 20th-century U.S. general). While not as much an inn as during Fitzhugh's time, Chatham continued a mecca for the Virginia gentry whenever the family was at home. They were often away; the horses and carriages of the Ellwood and Chatham families pounded the stones of the Orange Turnpike deeper into the clay in many a journey to and fro. The Judge's daughter Elizabeth Tucker Coalter on one visit to Ellwood received the thrill of her enthusiatic young life, meeting there General Lafayette on August 15, 1825. He had passed that way the year before, going east in public procession during the course of his triumphal tour of America. This time he was privately travelling west. He breakfasted and stopped a couple of hours. (Martha Jones happened to be there also.) Elizabeth Coalter sat by in a state of exalted veneration, grieving to think she would never again see this good old gentleman. He seemed so tired.

William Jones, a short, powerful figure in his knee breeches and ruffled shirt (which must have carried Lafayette's thoughts backward), was not young either, but he was not tired. His first wife had died in 1823. To the family's dismay, he married her grandniece in 1828, he then being 78 and the bride 16. A child, Betty Churchill Jones, was born in 1829, with, it might be said, an antique silver spoon in her mouth. At any rate, she rode in a coach and four; her daddy was reputed to have been the owner of the last such equipage in Virginia.

At Ellwood and Chatham, Janet Williamson, treated kindly by all, fitted in as best she might, while Elizabeth Coalter dazzled beaux and pleased even other belles. Her half uncle John Randolph of Roanoke wrote to her: "I love you more than worlds." This statesman, publicly known more for vituperation than expressions of affection, had lost touch with the Coalters until she wrote to him; then he corresponded with her and visited Chatham. Despite a poor opinion of marriage, he seemed a matchmaker too. He gave his godson John R. Bryan a letter of introduction to Elizabeth. Bryan, visiting Chatham, found that the family was at Ellwood. Persevering, he turned his horse's head west to that place. There he, too, found Elizabeth a fine girl. She, who once burst out,"Oh, how I love my kin," soon came to love Randolph's godson.

Little Betty Jones

Elizabeth's brother St. George Tucker Coalter, like John Randolph a consumptive, was as artless as his uncle was saturnine. Known to the family as Saint, St. George radiated cheer. "My wild Boy," his Grandfather Tucker called this madcap student of Mr. Jefferson's brand-new University. He was always, said his sister, injuring himself or breaking some bone or other.

St. George Coalter married a Miss Judith Harrison Tomlin in King William County and had six children in ten years. The last part of that time he lived at his own place, St. George's Park, behind Chatham, on land deeded to him by his father. Saint's cheerful resolution fought his increasingly poor health. He tried not to give up the parties he so loved, as those of November 1837, when he exulted in a great marriage party at Kenmore, the Gordon mansion in Fredericksburg, and a big treat given by his Mama Coalter at Chatham, where, in his words, "the whole nation" helped him enjoy "a grand splurge."

Saint was hectic. His indulgent father was staid. Victorian comfort enveloped Chatham. The retired Judge basked in a house of content. During the 1830's Charles Augustus Murray, grandson of Lord Dunmore (of unhappy memory in Virginia),* visited Fredericksburg. He liked the town and Falmouth and admired the situation of Chatham. Walking up and getting permission, the family being away, he toured the grounds. The house itself, the terraced garden, the straight green walks, the Lombardy poplars lined up like grenadiers—"all betokened the Old Dominion" to Murray.

Judge Coalter returned while the visitor was still there and insisted he come in for a glass of Madeira. Murray found Judge Coalter plain, courteous and hospitable, a sterling gentleman who had risen from no fortune to the highest rank by industry and integrity. The Judge was ever mindful of his good fortune. Once Elizabeth saw him moved to tears by a book. He told her it was about a poor man who rose to wealth, and it reminded him of his youth and the young wife, so soon dead, of his days of poverty.

Now, free of both poverty and the law, no longer jolted by a hard-trotting horse on circuit, he nodded over his books amid family and dependents; one of the latter, "Uncle Charles," acted the role of autocratic butler who regarded the place as his. And all went spruce, dressed in cloth of plaid and sober black from the spinning and weaving house, where the work chants kept time with the whir of the wheel and the clack of the loom.

*Among other misdeeds, Virginia's last Royal Governor burned and pillaged in a 1776 foray up the Potomac.

At age 68, Judge Coalter left all this in February of 1838 by the course of a paralysis that had come for him twelve months earlier. During that year, Saint suffered in consumption's later stages a weariness of the flesh. The heart, now sad, was on its last mile, though the body travelled farther, seeking health by change to King William County and to the western springs. In 1839, Saint came home to his little house on St. George's Park to die.

Elizabeth Coalter had married John Randolph Bryan in 1830 at Chatham and now lived at Eagle Point in Gloucester County. As Saint's end neared, a relative noted that Elizabeth was to have come up, but he supposed her husband's pursuit of money had somehow interfered. She did come up and she saw Saint die, calm and happy (the mourners said) in the heat of August. Sadly the sister went on the steamboat down the winding Rappahannock to the Bay and on to Eagle Point.

Saint's widow and children moved away, but Chatham did not lack juvenile visitors. Little Betty Jones of Ellwood began spending the winters with her half sister Hannah, old enough to be her grandmother, to attend school in Fredericksburg. Elizabeth Coalter Bryan's first child had been born at Chatham; she continued to return for a number of her ten confinements. She considered her Mama Hannah an extraordinary nurse who could, if all failed, always put on a black cap and rival the Sisters of Charity, whose fame as nurses took a local lady all the way to Baltimore for a delivery. When nurse Mama fell low to a stomach disorder, the family marvelled. She had inherited her father William's good digestion, which held up in his age, though he had to pamper it on occasion; visiting Chatham for Christmas of 1840, the ninety-year-old gourmet felt queasy owing to a large quantity of fried oysters consumed at supper and had to get up in the night to keep all down with pound cake. He continued to live well, dying at Ellwood in 1845.

Betty Jones, grown to beauty, married James Horace Lacy, a Presbyterian minister's son, at Ellwood in 1848. The life interest of William Jones' widow, to be traded for an annuity in the event of her remarriage, ended when she married in 1847. So the young Mrs. Lacy became mistress of Ellwood.

Hannah Coalter, carrying on at Chatham, seemed destined to rival her father in zestful longevity. In 1851, at age 71, she bought a fine pair of young horses and considered buying a new carriage of the latest style. However fast she rode, she did not equal her father in staying power. Thirty-one years mistress of Chatham, she died there in 1857 in her seventy-seventh year. She outlasted her stepdaughter; Elizabeth Bryan had died in 1856.

Hannah's will left all estate, real and personal, to Janet Williamson for life, exclusive of slaves and legacies. At Janet's death, all would go to Betty Lacy's children. The lordly butler Charles was set free with an annuity, and the other slaves (about 90) were also to be manumitted.

Janet certainly could not administer Chatham. She could not even stay there without family members. Her condition was worsening. The executors ordered a public sale. J. Horace Lacy bought the place, including the mill on Claiborne's Run, in November of 1857 for $36,950. He thought he could meet the payments "with Miss Janet's aid." It was arranged for the freed Charles to stay

Mrs. Betty Jones Lacy

J. Horace Lacy

on as the Lacys' major-domo. The other slaves, it turned out, remained as slaves. A legal technicality voided their manumission. Janet Williamson died at Chatham, December 6, 1858.

James Horace Lacy attended Washington College (later Washington and Lee University), Lexington, Virginia. He had been tutor to Elizabeth Coalter Bryan's children at Eagle Point and there met Betty Jones. Such were the coincidences of interconnecting families. Now he and Betty were master and mistress of both places of her youth. Quite a happy few—mother, father, Agnes, William, Elizabeth, Graham, and baby Lucy—travelled back and forth between a hill in the Wilderness and the turfed slopes of Chatham at the eve of civil war.*

*In 1861, Horace was 38, Betty 31, Agnes 9, William 8, Elizabeth 6, Graham 4, and Lucy 1.

"In the hands of the Philistines"

1. Fringe of War

"HERE WE ARE ON TOM TIDDLER'S GROUND picking up gold and silver," children used to sing in an old game as they converged on the goal keeper. Tom Tiddler's ground came to mean a place of good fortune, also debatable land. During the early part of a new decade the little Lacys in their pinafores and baggy pants left their play on Tom Tiddler's ground and crossed the Rappahannock, not on a routine visit to Ellwood, but to enter a new life. The Yankees were coming.

The short way overland to Fredericksburg from the Potomac, twice ignored by the British, appealed to the Federal forces of the American Civil War. That route became significant when the seceded states made Richmond, only fifty-five miles south of Fredericksburg, the capital of their Confederacy. The landing at the mouth of Aquia Creek, fifteen miles east of Fredericksburg, was the terminus of the Richmond, Fredericksburg and Potomac Railroad. Before the war steamboats conveyed travellers along the forty-mile north-south stretch of the river between Aquia Landing and Washington. Now Aquia and Belle Plain on Potomac Creek just downriver presented suitable bases for a move on Richmond.

The U.S. Navy's Potomac Flotilla cruised in to bombard Aquia during the last of May and June 1, 1861. Daniel Ruggles, an Old Regular from Massachusetts bound to the South by marriage, commanded the Department of Fredericksburg, whose troops beat off the attackers. J. Horace Lacy served him as aide-de-camp and also his successor, T. H. Holmes.

On another invasion route a Federal army under Irvin McDowell failed in July at Manassas, or, Bull Run. President Abraham Lincoln gave command to George B. McClellan, demoting McDowell to corps commander.

In the spring of 1862, McClellan transported his army down the Potomac and Chesapeake Bay to march against Richmond up the Peninsula between the York and James rivers. Lincoln, ever anxious to keep Washington well covered, held back McDowell's Corps. Seeking both to protect Washington and to help McClellan, McDowell brought his troops southward toward Fredericksburg, with the consent of the War Department, which, however, informed the Commander of the Department of the Rappahannock (McDowell's new title) that he could occupy Fredericksburg, but not with a view to further movement.

Inasmuch as Confederate General Joseph E. Johnston had fallen back from the Bull Run-Potomac line, McDowell encountered no significant resistance, though the vanguard of Rufus King's Division, marching down the Warrenton Road from Northern Virginia and reaching Falmouth and Stafford Heights on April 18, failed to stop the Rebels from burning the three bridges to Fredericksburg: the Falmouth Bridge, the Chatham Bridge and the Railroad Bridge.

On April 19, the guns of John Gibbon, a noted artillerist in the Old Army, overawed the town, having lobbed in a few shells during the skirmishing. Fredericksburg's Mayor Montgomery Slaughter and a committee met brigade commander Christopher Augur at Falmouth, assured him that Fredericksburg no longer harbored Confederate troops and requested protection under the rules of civilized warfare. The delegation was told that other units would soon be along. A committee member, Thomas B. Barton, Commonwealth's Attorney and the oldest lawyer in town, passed the information to the Confederate military. He also reported that the Presbyterian minister Beverley Tucker Lacy (Horace's older brother) had learned that a large force on the way to Richmond occupied Stafford. These intelligences accorded with McDowell's program of unsettling the enemy.

"We are in the hands of the Philistines," wrote the venerable Barton.

About 4 p.m., April 22, McDowell's wagon train arrived "on the camping ground at Mr. Lacy's farm, opposite Fredericksburg," a staff officer recorded in the headquarters journal.

The sharp hooves of the mules and the iron-bound wheels of the covered wagons cut new paths in Betty Lacy's lawns, and the afternoon light fell on a house of changing fortunes. Certain architectural changes already obtained. Nineteenth-century America liked porches; the Yankees could feel at home with several additions to the plain front of Fitzhugh's Chatham. In the middle of the house the gable roof of a two-storied verandah joined the main roof at right angles. Gable-roofed smaller porches covering the doors of each wing were connected to the ground floor of the verandah by unroofed walkways.

There was also a porch over the center of the rear, or east face, of the house. Eastward, winter wheat shone green on the fields of an abandoned plantation.

In these April days, often wet, the weatherbeaten McDowell, astride one of his horses or seated in his new carriage behind four dappled grays, kept busy inspecting supply lines and bases. Herman Haupt, a fellow West Pointer and more recently a railroad tycoon, set to work replacing the Aquia wharf, several miles of track and the bridges destroyed by the Confederates. Haupt used some fugitive slaves, or, "contrabands,"* but found them hard to retain, since "they were all bound for Washington, as they said, to see 'Massa Lincoln.'" Maybe some of them saw the President at Aquia when he met McDowell there on Sunday, April 20; with other business, Lincoln agreed to the promotion of John Gibbon.

Pending completion of railroad repairs, army wagons supplied the troops from wharfage rigged up at Belle Plain.

Though McDowell regarded Chatham his headquarters from the beginning, for a time he worked out of the Hedgman House on the railroad, two and a half miles from Aquia Landing, where the Fredericksburg elements of the 30th Virginia Regiment had camped in 1861. (This antique place was not to outlast the war.) While at Hedgman's, McDowell learned of sick being housed at Chatham; he informed Rufus King that Chatham was Department Headquarters, with some staff present, not a hospital.

Curious visitors, foreign and domestic, soon began descending on the occupied territory. McDowell interrupted overseeing the rail work to conduct Lord Frederick Paulet, commander of the garrison at Montreal, and party over to Fredericksburg. They also saw a troop review at King's headquarters, the Phillips House, behind Chatham.

McDowell had requested a steam ferryboat with lumber, as well as India rubber pontoons and some canal boats for bridging the Rappahannock. The twenty-boat canal-barge bridge, just below the ruined Railroad Bridge, allowed easy access to Fredericksburg on May 3 when McDowell was host to Secretary of War Edwin M. Stanton, Secretary of the Treasury Salmon P. Chase, Assistant Secretary of the Treasury George Harrington, and Congressman

*Union slang. Fugitive slaves within Union lines were considered contraband of war and no longer property of Confederate masters.

A. Lincoln.
Steel Engraving (1862).
Depiction by Alonzo Chappel

James K. Moorhead of Pennsylvania. The blacks of the town showed delight; other residents turned away or slammed shutters.

Several days earlier McDowell had ordered all officers not of his staff to leave the Lacy House and to remove their horses from the stables there. On busy May 3 he fully took over Chatham, but did not sleep there that night, feeling obliged to conduct his vistors back to Aquia. Neither did his staff sleep *in* Chatham that or following nights; the General reserved the house for himself. Wagons from Aquia arrived late; the officers retired supperless to tents.

On May 4, the engineers finished laying a pontoon bridge a little north of Chatham house, enabling the now more settled staffers to get good fresh bread expeditiously from a Fredericksburg bakery. The new crossing also served the purposes, whatever they were, of a May 4 visitor to Chatham, "a colored man . . .

representing himself as Jeff. Davis' coachman," who seemed authentic to McDowell. The coachman reported various conversations overheard; Mrs. Davis had said the Confederacy was "about played out." The Davis family, ready to move, was all packed up: books, clothing and pictures.

As May passed softly, Chatham attained the status of post office for the Department of the Rappahannock. Officers raced horses. John Gibbon presented himself at headquarters in the new uniform of an infantry brigadier. Other officers and their visiting ladies also came to call. Bands played. The usurpers relaxed in the balmy air and watched the sun set behind the steepled town.

On the warm Saturday of May 10, Betty Lacy and children, currently staying with friends in town while Horace campaigned on the Peninsula, trooped across the pontoons to see how their dwelling fared. Mrs. Lacy thought too many people were about, but found no great damage to the house itself. McDowell was an occupier with a sense of propriety.

In Fredericksburg, equally proper brigade commander Marsena Rudolph Patrick, the acting military governor of the town, dispensed even-handed justice from his office at the Farmer's Bank.* McDowell, himself sometimes surfeited with visitors, could himself be boring when visiting the bank. "He hangs on terribly," Patrick confided to his diary.

Both Patrick and McDowell tried to be correct and maintain occupation discipline. They deplored the hostile attitude of political officers. Patrick did not condone Secession, but remained gallant to "Secesh" ladies, though not partial to the arrogant ones, whom he found pious and insolent. McDowell, as commanding officer, bore assaults from all sides. One citizen, a Northern man living in Fredericksburg, kept waylaying McDowell at Chatham: in the hall, on the west portico, and at the back steps. He had axes to grind before he rested, and he considered McDowell's indifference to him treasonable. The General later remembered him simply as "a man annoying me with what seemed far more closely connected with his personal matters and his animosities than with the public service." Thus McDowell testified in an 1862-63 Court of Inquiry requested by himself to clear his name besmirched by gossip. The ruddy-faced General denied many allegations, among them suspicion of drunkenness and the charge that he had acted disloyally in saving the Chatham wheat from destruction; in fact, he saved it for Union use; it was ground in the old mill on Claiborne's Run. The Court in clearing him called the drunkenness charge ridiculous; unbiased witnesses testified he was notorious in the army for sobriety.

*NW corner of Princess Anne and George streets, since the Civil War occupied by the National Bank of Fredericksburg.

(Continued on page 49)

Section Showing Early Fitzhugh Plantation at the Site of Future Chatham.

"A MAP *of* THE NORTHERN NECK *of* VIRGINIA, According to an Actual Survey Begun in the Year 1736 and Ended in the Year 1746, Drawn by Peter Jefferson and Robert Brooke."
From a tracing made for the Commonwealth of Virginia in 1860 of the original in the British Public Record Office.
Courtesy of Virginia State Library

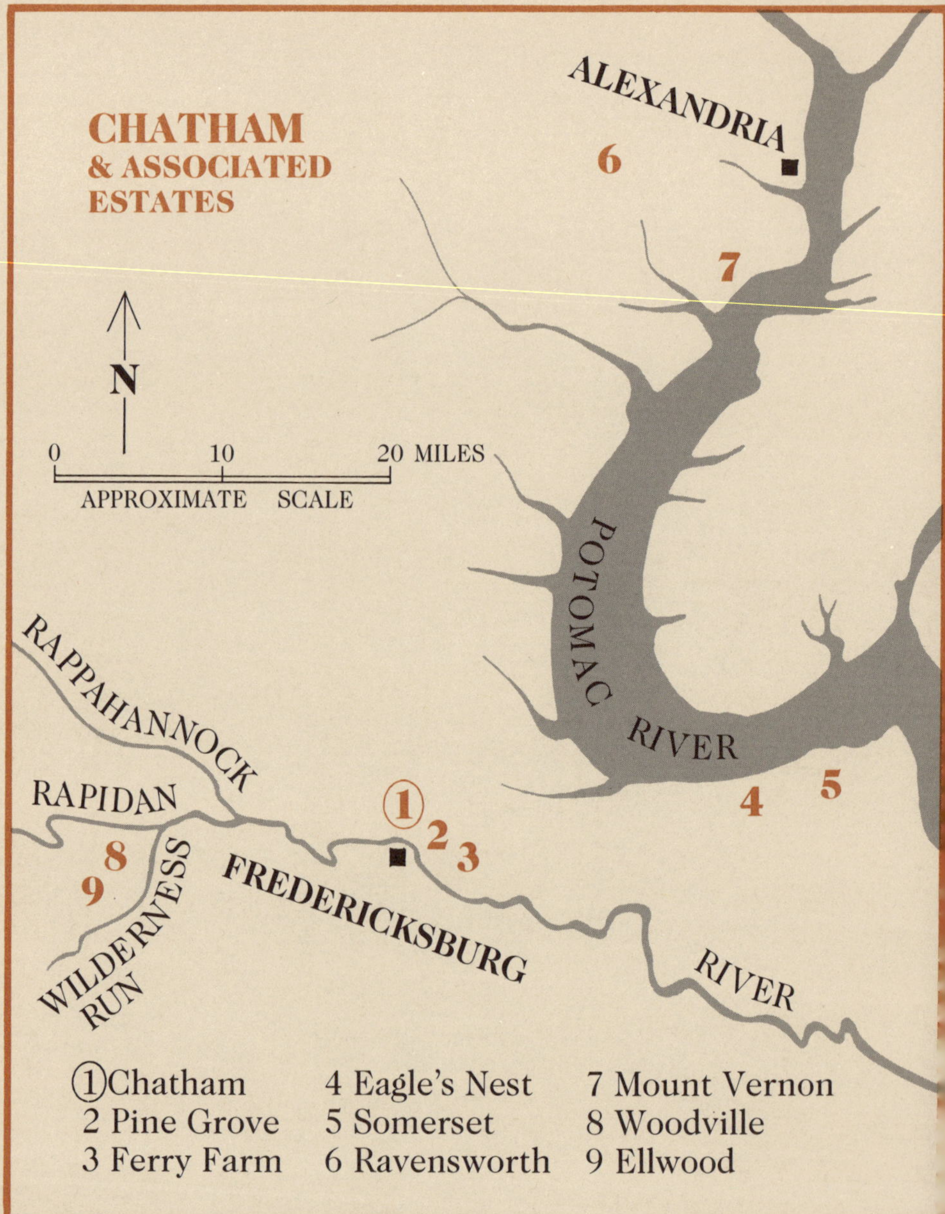

CHATHAM
& ASSOCIATED
ESTATES
ALEXANDRIA
N
0
10
20 MILES
APPROXIMATE SCALE
POTOMAC
RIVER
RAPPAHANNOCK
RAPIDAN
WILDERNESS
RUN
FREDERICKSBURG
RIVER
①Chatham
2 Pine Grove
3 Ferry Farm
4 Eagle's Nest
5 Somerset
6 Ravensworth
7 Mount Vernon
8 Woodville
9 Ellwood

EASTERN THEATER *of* THE CIVIL WAR.

From a Park pamphlet of the 1930's, drawn by Neil C. Goodman

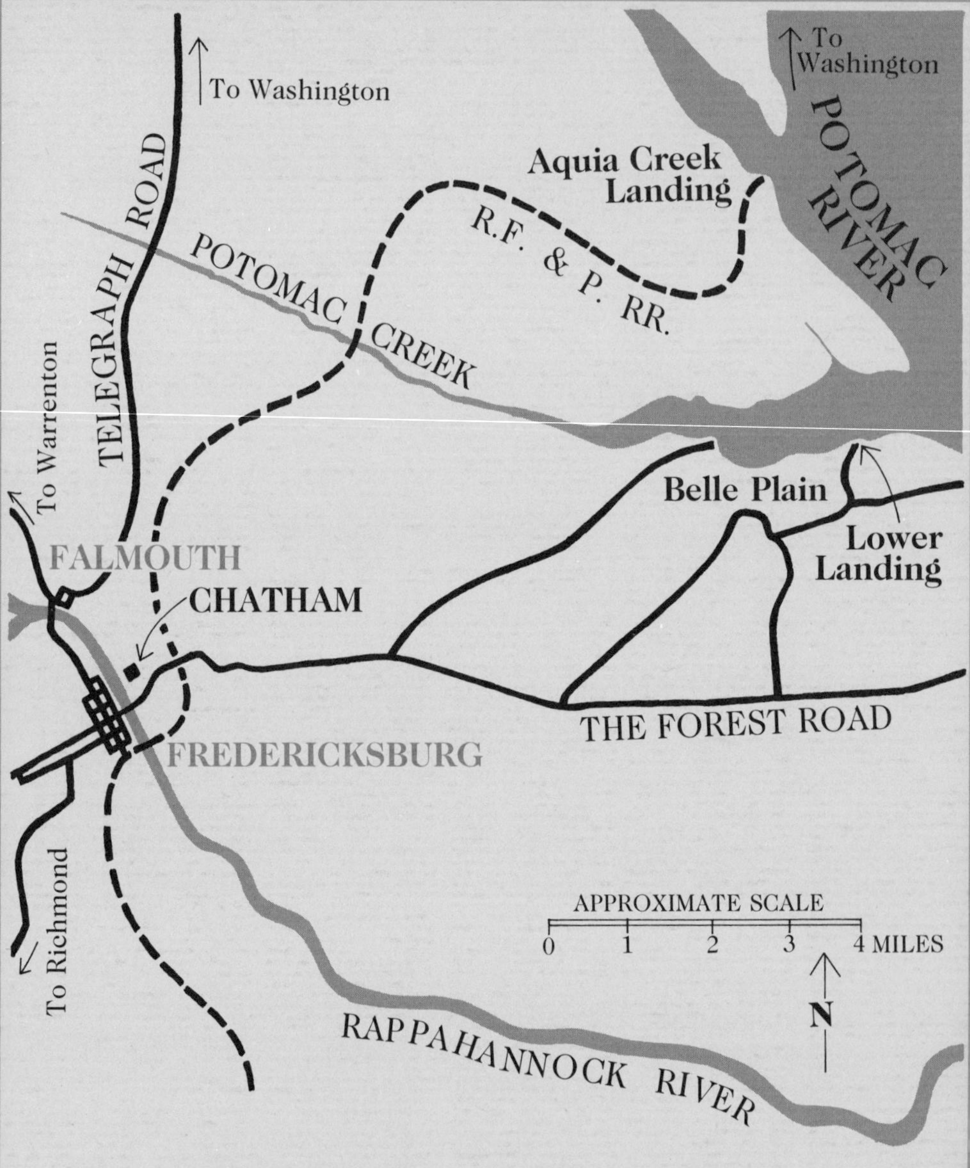

UNION COMMUNICATIONS

To Warrenton
To Washington
To Aquia & Washington
FALMOUTH
DAM
ylor
CANAL
S
Lacy
B
Phillips
G
Wallace
King
To Potomac Ck.
ORANGE PIKE
Marye
HAZEL RUN
N
0
½
1 MILES
APPROXIMATE SCALE
Howison
TELEGRAPH ROAD
To Richmond
R.F. & P. RR.
DEEP RUN
OLD RICHMOND STAGE ROAD
RAPPAHANNOCK RIVER
BATTLE of FREDERICKSBURG
December 11-16, 1862
Main Union Attacks, Dec. 13
Union Retreat, Dec. 15-16
S Sumner's Headquarters, Dec. 13
B Burnside's Headquarters, Dec. 13
G Gen. Hdqs.—Before & After
Lee's Battle Station
National Cemetery, Postwar
Hamilton
HAMILTON'S CROSSING

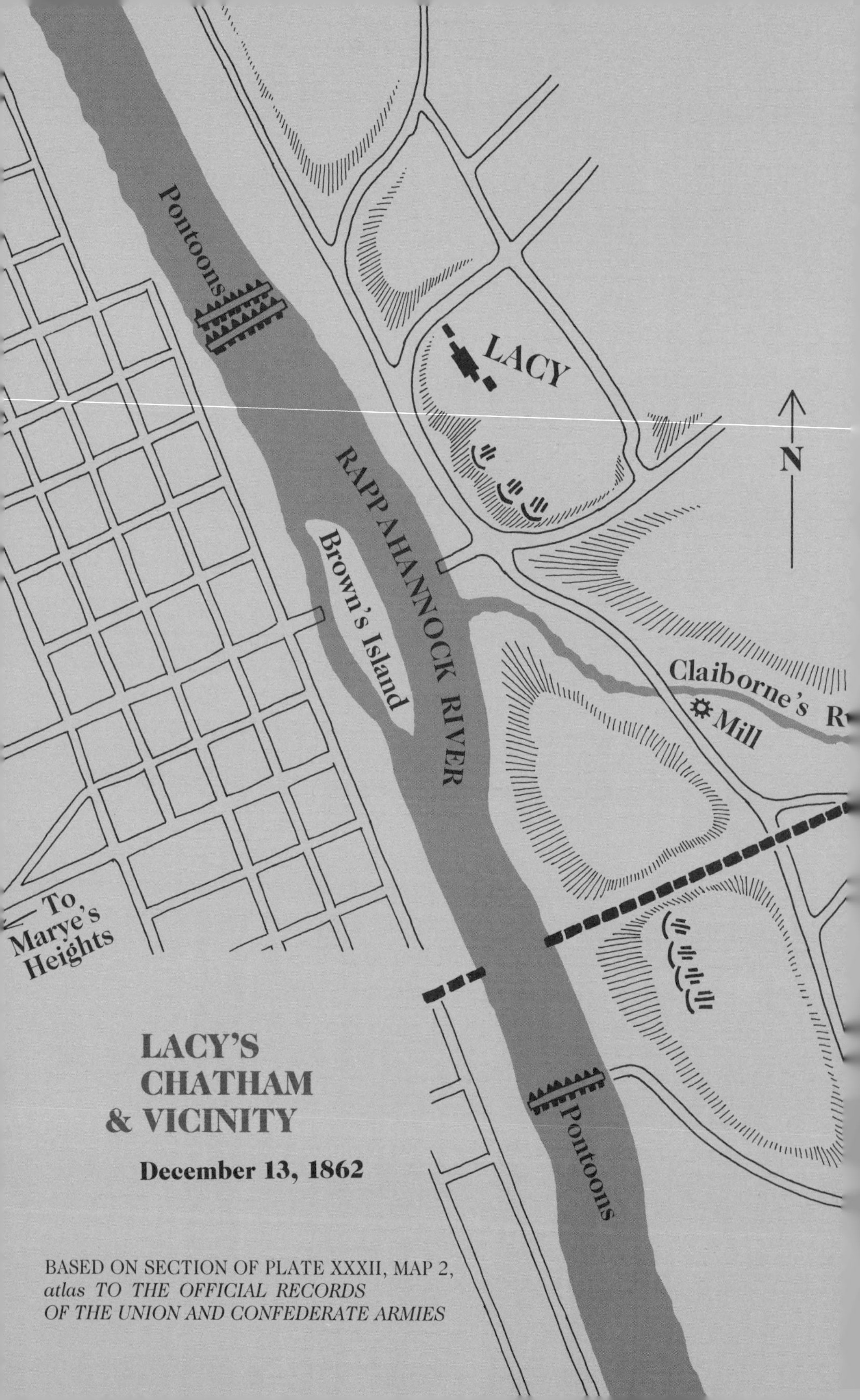

LACY'S CHATHAM & VICINITY

December 13, 1862

BASED ON SECTION OF PLATE XXXII, MAP 2, *atlas TO THE OFFICIAL RECORDS OF THE UNION AND CONFEDERATE ARMIES*

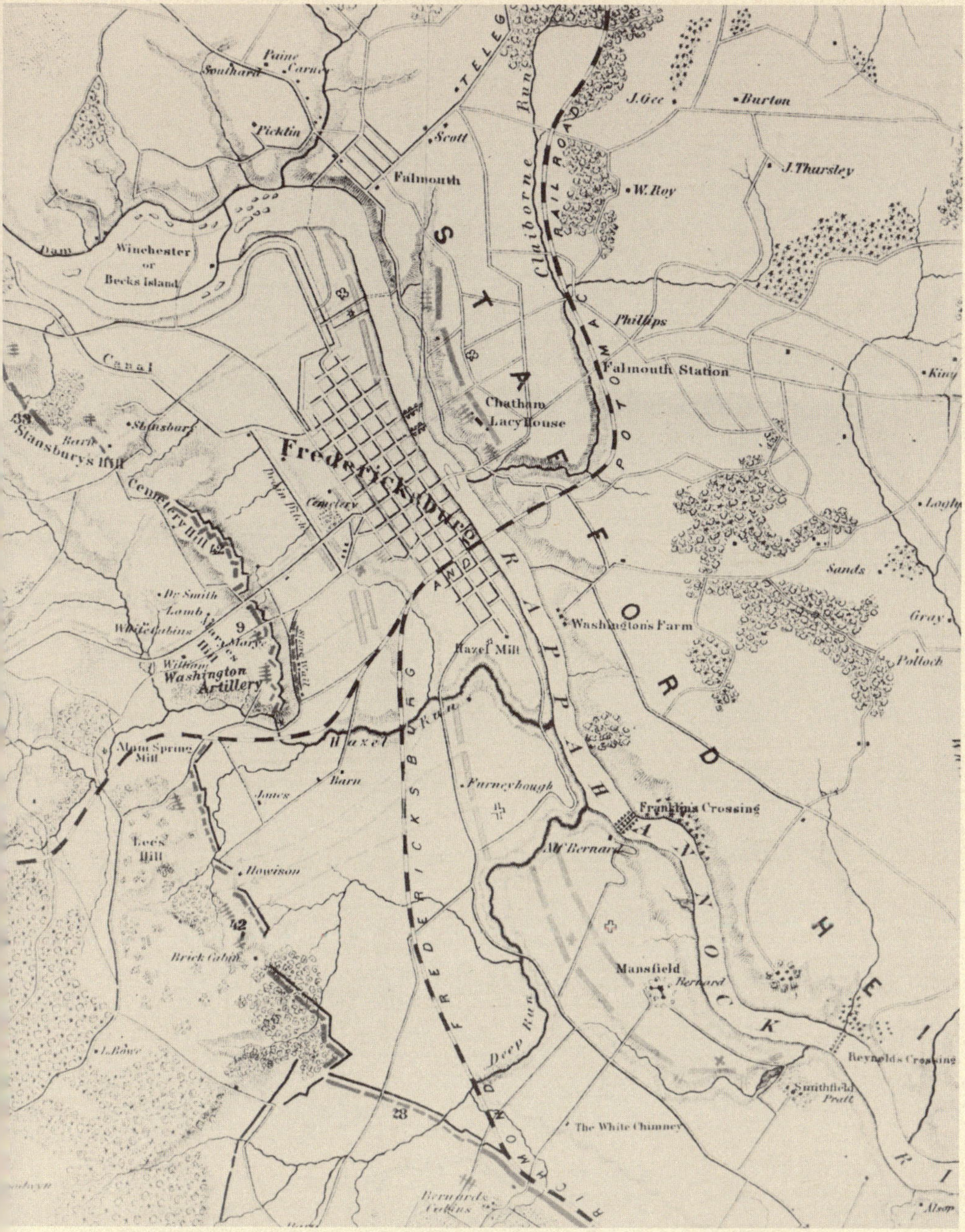

Section of Jed. Hotchkiss' map of **THE FIELD *of* SECOND FREDERICKSBURG, 1863.**

From *Chancellorsville* by Hotchkiss and William Allan, 1867

CHATHAM ESTATE

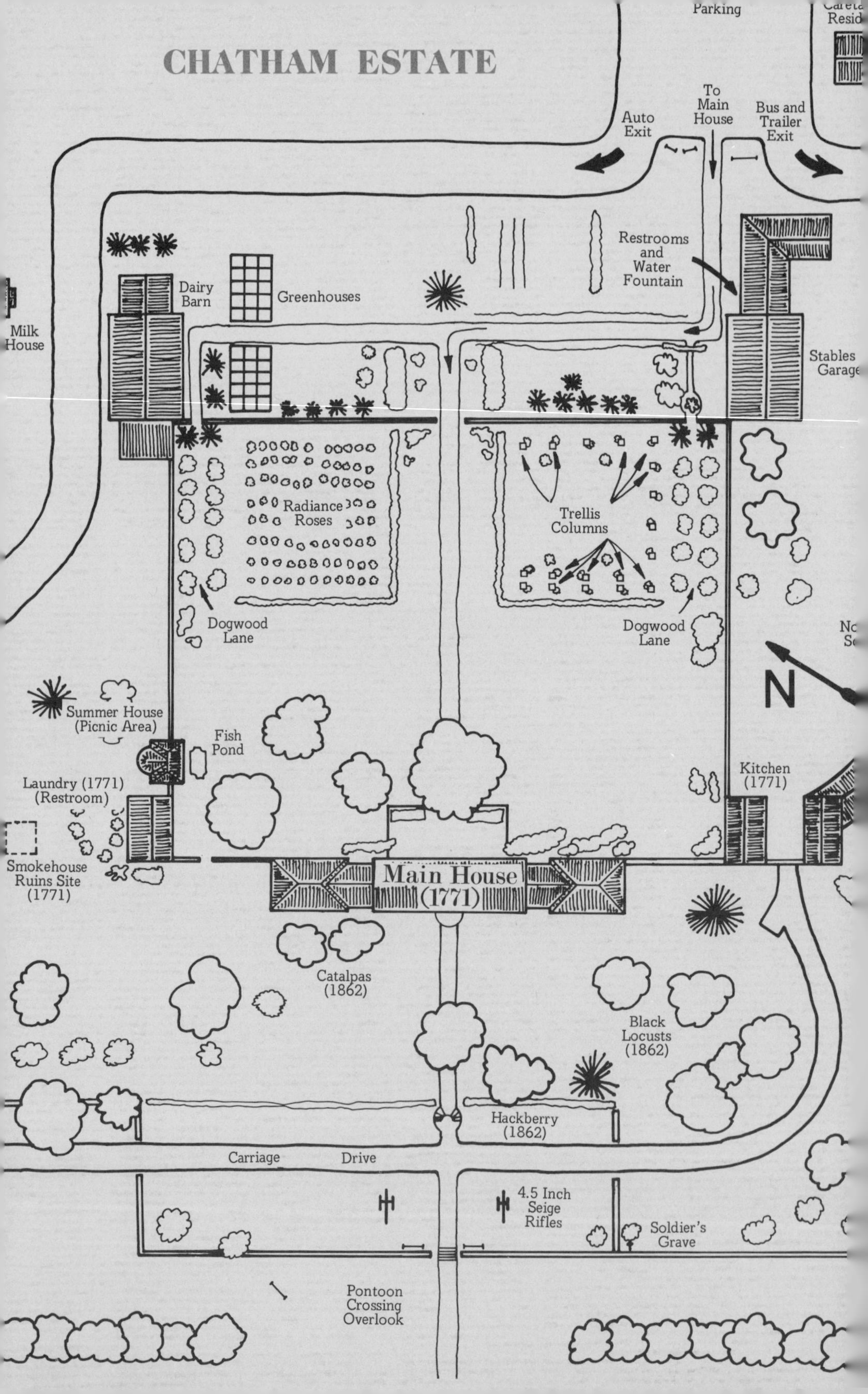

An occasional encounter with Rebel outposts beyond Fredericksburg gave spice to the spring of 1862, but mostly, except for Haupt's men, the Department of the Rappahannock was like McDowell at the bank—hanging about. By the middle of May, however, the Administration began to think a move on Richmond from Fredericksburg might be a capital idea. On the 17th, Secretary of War Stanton gave the order to march as soon as McDowell could be reinforced with another division. By Monday, May 19, the trains on the Aquia railroad were running into Fredericksburg. The supporting division arrived May 22 and went into camp near the Phillips House, the commander reporting in person at Chatham. McDowell, with more than 40,000 men and more than 100 pieces of artillery, planned to start May 25.

On Friday, May 23, President Lincoln, Secretary Stanton, the French Minister Henri Mercier, and Commander John Dahlgren came rolling in on Haupt's railroad. That morning McDowell conducted the party to see Patrick at the bank in Fredericksburg. McDowell's carriage and the four grays provided a fitting equipage for the President. Patrick sensed a feeling of sadness in the town and felt sad himself to think its helpless families, after he and most of McDowell's force marched away, would be left to the mercies of politically minded officers.

On the return to Stafford the visitors may well have been treated to a gourmet lunch at Chatham. At any rate, McDowell's two French cooks, who had been unhurt when their wagon overturned on the way down from Northern Virginia, were presumably in business at headquarters. During the afternoon, amid the village of tents and outbuildings given over to military business, officers crowded the grounds to meet Mr. Lincoln. The President held forth in Chatham's hall with his customary humor. He asked Gibbon if he were the man who wrote *The Decline and Fall of the Roman Empire*. When the ex-artillerist seemed at a loss how to reply, Lincoln went on to say, "Never mind, General, if you will write the decline and fall of this rebellion, I will let you off." Gibbon explained that he had written an artillery manual.

After the reception, Lincoln reviewed troops. It was a time of pleasantries and pageantry. That Friday evening McDowell accompanied the Presidential party back to Aquia. Lincoln suggested that McDowell not move toward Richmond until Monday.

On Sunday, McDowell received word from the President to lay aside the movement on Richmond and to put his men in motion toward the Shenandoah. Confederate General Thomas J.

("Stonewall") Jackson's Valley Campaign had rekindled the Administration's fears for the safety of the National Capital. Haupt, called to Chatham, found McDowell stewing, asserting that Washington was in no danger and that missing this chance against Richmond would prolong the war indefinitely.

The last dignitaries to visit Chatham that season, Chase and Harrington, birds of ill-omen redefining Mr. Lincoln's cheeriness, helped the General pass the Sabbath. As usual, he accompanied his guests back to Aquia.

Stanton urged all speed westward. McDowell and part of his staff departed Chatham. On the 27th all headquarters personnel left for Aquia. By midnight everything had been stowed aboard the steamboat *North America,* which carried a full load, including fugitive slaves "of all classes, sizes, and ages."

On May 31-June 1, Johnston engaged McClellan at Seven Pines, almost in Richmond. Upon Johnston's wounding, Robert E. Lee assumed command of the army he was to immortalize.

Horace Lacy, now a volunteer aide on the staff of Gustavus W. Smith, commanding the Confederate left wing at Seven Pines, escaped wounds or capture, only to be taken by Federal

Ruined Chatham Bridge, 1862. From a Civil War Souvenir Envelope

pickets while visiting the Fredericksburg area.

Seven was a lucky number on the Peninsula for the Confederates. Lee proceeded to save Richmond in the Seven Days' Battles. McClellan encamped down the James. After McClellan's failure, Lincoln shelved him. A new scratch army came into play in Northern Virginia under John Pope.

During the maneuvers of early summer, King came back to Stafford Heights. He established headquarters at Chatham. Whatever was expected of the future, for a time exuberance reigned. The solid old flooring that had recently felt only the decorous tread of dismounted officers and orderlies now resounded to the clatter of hooves. An old servant left by Mrs. Lacy to look after things—and *look* was about all he could do—often saw young officers ride up the front steps, through the hall, and out the back. Doubtless they refrained on July 12 when McDowell, on a visit to his old headquarters, sent for Patrick to talk over the military situation and complain of his troubles.

Patrick, back for another stay in the area, had been annoyed to find no bridges, only a ferry to Fredericksburg. A rainy spring was ushering in a wet summer. On June 4, the river had risen so high that officers in charge had decided to take up the Chatham pontoons. That afternoon a trestle bridge below the old Chatham Bridge broke up and carried off the repair work of the Railroad Bridge, the combined debris then smashing the canal-boat bridge. High water always draws riparian dwellers. A crowd along the Fredericksburg shore cheered and waved handkerchiefs as each structure gave way. Gunboats downstream managed to save the canal boats for putting in place again. A wire suspension bridge was built to Chatham. The Railroad Bridge, again repaired, washed out twice more that summer.

In early August, Patrick and King left the area to cooperate with Pope. Ambrose E. Burnside, following successes on the North Carolina coast, arrived to hold the Rappahannock. Part of his Ninth Corps went to Pope. Placing troops on both sides of the river, Burnside pitched his tent at Chatham. He was in good cheer; the public acclaimed him; and his servant and confidant Robert Holloway, captured at Bull Run, had been exchanged and returned to him. This faithful follower had been with him since the Mexican War. By general orders of August 12, Headquarters Ninth Army Corps, near Fredericksburg, Va., Burnside prohibited plundering. He, too, believed in carrying on war humanely, but civilians were sometimes guilty of inhumanity to each other.

General Marsena R. Patrick.
Photograph Courtesy of the National Archives

General Irvin McDowell.
Photograph Courtesy of the National Archives

Shortly before leaving, Military Governor Patrick had become caught up in a matter of hostages. At an earlier period, certain Fredericksburgers had caused several Northern-born citizens of the area and other presumed disloyal men, including a particularly cantankerous native, to be arrested and sent south. Then the U.S. Government proceeded to arrest a number of Fredericksburg men in retaliation. On July 23, four, including Thomas Barton, were conducted north, "greatly," observed Patrick, "to the disgust of Barton." Because of age and illness, however, the old burgher was paroled to friends in Baltimore.

More arrests followed. On August 13 a last batch of citizens was gathered in, among them Mayor Slaughter and George Henry Clay Rowe, who recorded his experiences in a diary. The next day on the way to imprisonment in Washington these men crossed the wire bridge to Chatham and met General Burnside. He received them courteously, wrote Rowe, dismissed their guard, gave them the freedom of the grounds and a stirrup cup of fine whiskey as they prepared to leave for Falmouth Station, the Federal depot

General Ambrose E. Burnside.
Steel Engraving (1862).
Depiction by Alonzo Chappel

General Edwin V. Sumner

between Chatham and the Phillips House. Ultimately, Confederate authorities decided that the people they had arrested were really inoffensive, which indeed they seem to have been, and everybody got released.

Burnside was soon to entertain fellow officers brought to his neighborhood by the military situation. McClellan's troops were coming up the Bay and the Potomac to reinforce Pope. Fitz-John Porter, who had distinguished himself on the Peninsula, arrived at Aquia, August 21, to bring his corps over the rails to the Rappahannock. He sent word to Burnside: "Will be at your place about dark, and want the shady side of a tree or house to-night."

On August 24, George Brinton McClellan himself reached Aquia. For the next couple of days "Mac" travelled back and forth to Chatham, sent Porter up the river and studied the ground with his old friend "Burn," from whose headquarters he assured the War Department of hearty co-operation. Then he left for Alexandria to dispatch more men to Pope. He lent Burnside his personal escort, a squadron of Regulars, "Burn" being weak in cavalry.

During the last days of August, Lee trounced Pope in the Second Battle of Manassas. Pope retired to the defenses of Washington. (McClellan wrote his wife from Alexandria: ". . . I have sent up every man I have . . . I have been listening to the sound of a great battle in the distance. My men engaged in it and I away!") Porter, accused of contributing to Pope's debacle, burned a long time in a heat worse than that of the unlucky day he landed at Aquia and went by rail to Chatham for the march upcountry. McDowell, too, suffered in the recriminations following Second Manassas; he had, in fact, never been forgiven for Bull Run, Manassas One.

In wild exodus from the Fredericksburg area, Burnside's people destroyed war material they might easily have saved. Smoke wreathed above the Fredericksburg Railroad Bridge, above Falmouth Station, above Aquia Landing. Along the Potomac green persimmons awaited frost. The last of the Philistines left Stafford County.

The River Crossing Near Chatham, December, 1862.
Lithograph of a Painting by Thure de Thulstrup

Lincoln offered McClellan another chance. McClellan muffed it at Sharpsburg on Antietam Creek, Maryland; he did stop Lee's invasion, but might have destroyed him. He followed Lee's retreat slowly. The country demanded action. A sad President presented the Army of the Potomac to Burnside. Burnside elected to move on Richmond by way of Fredericksburg. So the Philistines returned. Haupt again repaired the railroad. Persimmons hung gold on the leafless trees to delight Yankee boys sick of hardtack and rusty bacon.

2. Battlefield & After

ON NOVEMBER 19, Burnside and Patrick, now Provost Marshal General of the Army of the Potomac, met at Chatham to consider the situation. On November 21, Patrick crossed the Rappahannock by boat under flag of truce to order surrender of the town under threat of bombardment. On the 22nd, he received Fredericksburg authorities and Confederate officers at Chatham. The Federals refrained from bombardment, though the town was not surrendered.

Meanwhile, Burnside sat like a broody hen. He had reached the Rappahannock before the pontoons so vital to his plans and thus shown his probable area of attack. The Confederates began arriving to oppose him. In late November the President steamed down the Potomac to meet and reassure a General under strain—and perhaps to reassure himself. Further conference took place in Washington.

The pontoons arrived before Lincoln's visit and before all of Lee's men arrived, but Burnside cogitated at the Wallace place, Little Whim, a mile east of the Phillips House. Finally, he decided to

Clara Barton.
Photograph Courtesy of the National Archives

strike; he would attack at and near Fredericksburg rather than points some distance above or below. Several generals of his right wing, commanded by Edwin V. Sumner, thought the heights behind Fredericksburg too formidable. Burnside evidently had not explored much across the river during his earlier Stafford stay. This dissent irritated the normally amiable Burnside. He asked the objectors to meet him the night of December 10 at the Lacy House. Discipline prevailed throughout this council. Everybody, disclaiming discourtesy, offered allegiance so solemnly an officer walking in late said he thought he was entering a Methodist camp meeting.

Three crossing sites were chosen: the old one just upriver from Chatham running to Fredericksburg's Hawke Street; a second, just below the railroad at the site of McDowell's canal-boat bridge; and a third, a mile or more below that.

Men, horses and guns swarmed over the Lacy plateau. Colonel Charles H. Tompkins, commanding the right center of a long line of cannon, posted guns some way off on either side of the house, his battle station. General Daniel P. Woodbury, commanding Engineer Brigade, took up headquarters there too.

Down from the Chatham ridge and the hills southward, the pontoon wagons rattled through the early morning mist of December 11 to unload at the river. Breaking though ice, boat by boat and plank by plank, the engineers sought the invisible enemy shore. Now and again a rifle shot spattered out of the chill vapor. Then the sun began to burn off the fog, and the minie balls sang in crescendo from houses, walls and barricades that were becoming ominously visible. Here and there an engineer fell; the rest fled. At the third site, the Confederates were soon driven off, but opposite the town William Barksdale's Mississippians stopped all work. The Union officers could not make their men step out. Woodbury led a group of volunteers to the Chatham shore. They balked too.

The sun shone for the Yankee artillerists as well as for Rebel riflemen. No longer firing blind, the gunners blasted the town. Burnside had spent most of the morning at Chatham and gone down to the river to view the impasse up close. He talked with Woodbury and Chief of Artillery Henry J. Hunt. Hunt suggested sending troops

Working the Beardslee Telegraph Machine.
From Beardslee's *Military Telegraph*, 1863

Civil War Chatham from N.W.
Photograph Courtesy Library of Congress

over in boats to gain a bridgehead. Hunt and Woodbury consulted Norman J. Hall, commanding a brigade at the Lacy House, about the practicality of the proposal. Hall agreed immediately. The 7th Michigan, Hall's old regiment, volunteered to go first. The Michiganders rushed to embark. Robert Henry Hendershot,* a drummer boy twelve years old that day, hopped in a boat. He was ordered out. He then helped push off, neglecting, however, to let go, enduring the icy waters to arrive and run up the other bank with the soldiers chasing the Confederates. An exploding shell broke his drum. He threw it aside, picked up a dead man's musket and joined the fight. The 19th and 20th Massachusetts followed to clear the town. At the middle site, the 89th New York also crossed in boats. The town bridges were then soon finished, but the day had gone.

Up at Chatham, Provost Marshal General Patrick sat watching it all. Clara Barton,** nurse and one-woman relief corps, looked on from the second story of the verandah. To the old maid it was gore and glory. She noted in front of Chatham the leader Burnside, "of honest heart and genial face." To Patrick, Burnside was merely another general. The flames Patrick saw rose from a town he had, in some sort, become a part of.

The Phillips House, Sumner's headquarters, high on the second tier of heights, dominated the field. From there a signal station communicated with stations below, one at the Lacy House.

*Hendershot's name has in some accounts been mistakenly written Henderson. A son changed his own name back to the original German: v. Haydershatte.

**Relationship of this Yankee lady to Thomas Barton of Fredericksburg is unknown. Both probably would repudiate any kinship.

Signal flags evoked past warfare; the wires of G. W. Beardslee's magnetic telegraph stretched over this battlefield like threads to a technocratic future. The inventor's son Frederick, recently appointed captain, oversaw the telegraph operations.

On the 12th, the bulk of the army crossed. Clara Barton went along from Chatham.

On the 13th, defeated at all points, the Federals met disaster in futile charges against Marye's Heights behind the town. Signal flags from the town's Court House steeple waved bad news to Chatham, and flags and wire carried it from there to the Phillips House. Actually, the messages did not bring out the horror, which could be seen plainly enough by looking through a large telescope at the Phillips House. Burnside occupied that house during the fight, and Sumner moved to Chatham. The white-bearded Sumner, an old

Military Cookhouse Built Against Chimney of a Ruined Outbuilding, Chatham. Contemporary Drawing

Indian fighter partial to close action, had established headquarters in town during the morning, but Burnside had requested return. The Lacy House was set as his limit. That was not beyond danger; Confederate cannon could just about reach it. Lieutenant David Wonderly superintended his telegraph station there under fire; one shell burst in the cellar.

The Federal attack force retreated to Stafford during the night of December 15-16, many of its members bringing back, despite Patrick's vigilance, a collection of loot. Other valuables littered the streets of the once well-off, now gutted, city. Sumner was again at the Phillips House and Miss Barton had returned to help the surgeons in a Chatham crammed with wounded and dying covering all available space of every room and even, as she never forgot, the floors of the frigid porches.

Later in the month, Walt Whitman visited Stafford. He had come to check on a wounded brother who, it happened, had not been badly hurt. He then looked around, seeing the war at first hand. At Chatham he saw the piles of amputated limbs common to Civil War hospitals. He saw the former pleasure ground here turning to graveyard. If the American Red Cross sprang from Clara Barton's wartime experiences, so did the sights of Chatham affect the sensibilities of "the good gray poet." From this grisly Lacy House he went forth to a commitment of comforting the wounded and dying and exalting the suffering of America in verse.

The new year brought a new leader, Joseph Hooker. Hooker, handsome, sociable, not addicted to temperance, began restoring Union morale. He set a good example by sleeping in a tent—"as a soldier ought," remarked a chaplain—in the general headquarters area beyond the Phillips House. He could see the conflagration of that Victorian mansion on Valentine's Day, 1863, if he were home that day rather than, possibly, up at Willard's Hotel in Washington.* Somebody had fired up a camp stove in the Phillips attic.

Kept from the front in December, Sumner, on his way to another assignment, died in sickbed in March.

Chatham, its woodwork defaced and partly removed, battered, bare, became a hangout for the squads coming off the picket line. Mrs. John Harris of the Ladies' Aid Society of Philadelphia, assisted by a Mrs. Beck and Mrs. Mary W. Lee, an immigrant from Northern Ireland, cooked for the boys there. Besides coffee, tea, bread, and stewed fruit, they dished up hot ginger panada, a

*He certainly was at that popular gathering place for officers and favorite haunt of his, Februry 23; a dispatch from camp is addressed to him there.

mixture of corn meal, ground ginger, crackers, and wine. These ladies not only nourished the shivering pickets; on occasion they assisted Provost Marshal General Patrick by searching female prisoners suspected of spying.

The spring of 1863 came hesitantly on—there was a snowstorm April 4. The new man on horseback, known as "Fighting Joe," studied war and movement. Lincoln came down to review the army. Hooker, grand in his uniform on his white charger, spoke of early arrival in Richmond. The stovepipe hat of the tall black-clad President reached skyward while his feet hung almost to the ground from a borrowed horse. He enjoyed seeing the soldiers, but heard the General with misgivings.

Through April's last days, part of Hooker's army marched up the river to turn Lee's flank; part stayed to attack Fredericksburg from Stafford. Lee foiled the turning column at Chancellorsville, an old tavern ten miles west of Fredericksburg. The town attackers broke through Marye's Heights and marched on to try to aid the Chancellorsville group, but were stopped halfway at Salem Church. Then Hooker ordered everybody back to Stafford. He had been outgeneraled.

John Gibbon, commanding a division, was supposed to co-operate with the Fredericksburg attackers. On May 2 he had headquarters at Chatham, where Lincoln had joshed him the year before about helping in the decline and fall of the rebellion. He crossed pontoons at the old Chatham site and then, May 3, failed to break through the defenses extending north of Marye's Heights. When Marye's Heights fell to the force he was trying to help and the Confederates moved back, he stayed behind to guard the town. He would uphold the President much better on later occasions.

Again the telegraph worked from Lacy's and cannon loomed atop the Chatham hill. Before the Federal group near Salem Church was cut off by Confederate recapture of Marye's Heights, wagon trains from Chatham supplied that force. The Salem Church and Chancellorsville units retreated over the upper river. Gibbon returned over the Chatham pontoons.

In June both armies left the area, and in July Lee suffered defeat at Gettysburg.

The Wilderness-Spotsylvania battles of May 1864 made the whole town of Fredericksburg a Federal hospital. The Irish Mary Lee, again with the army, could find no stove to cook on. She remembered the one at Chatham, went over, found it still there, and had it brought to town.

During 1864-1865, the County Court met at Chatham. Battered the Lacy House may have been, but it was apparently more cheerful and handier than the official structure, called by a York Stater "a tumble-down and filthy building," at the village of Stafford Court House. Union soldiers had not contributed to neatness by scattering and looting records dating back to the middle 1600's.

In 1863 and 1864 the war reached Ellwood, the Lacy House in the Wilderness. The Confederates established a field hospital there in the Chancellorsville fighting at the height of Southern promise. Federal generals gathered there in the Battle of the Wilderness on the way to final Northern victory. Toward the end of the war, Betty Lacy and children sought shelter in Southwest Virginia, where her husband, long since exchanged, was stationed. The culmination of his military career was the position, ranking as major, of chief inspector of field transportation of the Trans-Mississippi Department.

Furniture from Chatham and Ellwood also found shelter in the western part of Virginia. At war's end it sank in a canal boat transporting it homeward, but was fished up and dried. Only the more valuable pieces had been moved in the first place. Such things as rocking horses were left behind; indeed, Lacy family archives mention no rocking horse. The quiet mount that touched the heart of the chaplain-reporter at Chatham in early April of 1863 appears and disappears with his sighting. It may not have been a Lacy mount, but rather one of so many things looted from Fredericksburg. In any event, for its brief moment, it bears witness, riderless.

Upon the Lacys' return east, they stayed a while at her mother's place near Ellwood. Ellwood was not then really habitable and was, anyway, for the time being, occupied by a squatter, a carpetbagger who refused to move. When the Major set about rehabilitating Chatham, the family introduced to the old home a new member, Sallie McGavock* Lacy, born in Southwest Virginia, April 3, 1865, the day Richmond fell. In time, she would see both Chatham and Ellwood restored to amenity. The bodies from the graves surrounding Chatham would be removed to the Fredericksburg National Cemetery, except for two or three overlooked, and the hummocky lawns leveled and reseeded. Flowers would bloom again. But the place would never be to the Lacys the fortunate ground of yore. To them and their peers of the South the past seemed to glitter brighter the farther it sped backward, a strange star defying the laws of optics.

*The name of the Pulaski County family who lent the refugee Lacys a house.

Private to National Estate

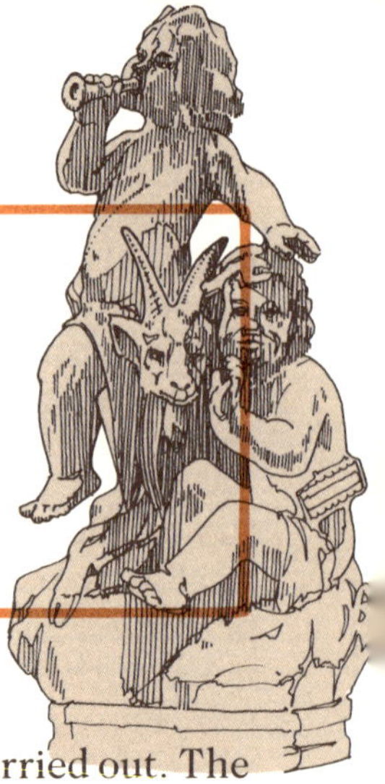

HANNAH COALTER'S WISHES had finally been carried out. The Chatham slaves were manumitted, though the census of 1870 shows that at least five, ranging in age from 14 to 80, had stayed on as servants. There were then two more Lacys: James Horace, Jr. and Beverley Randolph Drury. Graham, a veteran of wartime adventures, was no longer the baby boy.

In the early years of peace, officers and men of the graves details were welcome enough at Chatham, but unbidden civilian visitors annoyed the Lacy family. Graham, often acting as guide and considering it entirely proper "to spoil the Egyptians," was in turn annoyed when his parents forbade acceptance of any gratuities. He once resentfully showed a couple a room they seemed to have heard of and was astonished when the woman burst into tears. Her son, she said, a mere boy himself, had died there.

Efforts to recoup fortune and re-establish two farms did not keep Major and Mrs. Lacy from engaging in public affairs. Mrs. Lacy helped organize and served as a vice-president of the Ladies' Memorial Association of Fredericksburg, one of the aims of which was establishment of a local Confederate cemetery. Major Lacy, an orator of the old school, travelled and made speeches to raise money for this endeavor.

The oldest Lacy child, Agnes, married at Chatham in April of 1871 the Reverend James Power Smith, minister of the Fredericksburg Presbyterian Church. Captain Smith, born and reared in the North, was attending Virginia's Union Theological Seminary, then located at Hampden-Sydney College in Farmville, when Virginia joined the Confederacy. He had, after prayerful consideration, chosen to go with the South, becoming an aide to General Thomas J. Jackson. Agnes' uncle, Beverley Tucker Lacy, the Fredericksburg Presbyterian supply pastor at war's beginning, became a chaplain with Jackson, virtually a member of the General's staff. In years to come, Ellwood, the Lacy House in the Wilderness, was to be a summer bivouac of Confederate remembrance. Discussion there led to a monument for "Stonewall." In the early '80's, the Lacy brothers and Smith marked the site of Jackson's mortal wounding on the Chancellorsville field with a quartz boulder, the beginning of a campaign for a monument, which was dedicated in 1888.

Before those days of reminiscing at Ellwood, Major Lacy relinquished Chatham, selling in 1871 this farm of 468½ acres and a 44-acre Claiborne's Run tract to Oliver Watson of Williamsport, Pennsylvania. Watson, a lawyer and a banker, owner of farms in Pennsylvania and Virginia, leased Chatham to Samuel Shadle, a Pennsylvania farmer who later moved west.

Following sale of Chatham, Major Lacy was elected to the state legislature from Spotsylvania County and served one term.

After Watson's death in 1882, the banker's Chatham heir, Emma Watson Jones, conveyed Chatham to her brother, Oliver Watson, Jr. This Watson and his wife occupied the dwelling and made themselves a part of the community. He served as an officer of the Fredericksburg Agricultural Fair Association, was one of the owners of the Rappahannock Electric Light and Power Company and a vestryman of Trinity Episcopal Church.

In 1886 he added horse racing at Chatham to the attraction of the annual Fredericksburg Fourth-of-July regatta, held that year on Monday, July 5. Tolls were waived on Scott's (Chatham) Bridge for the event. The beflounced and parasoled ladies and the linen-suited sports rolled in surreys and buggies up the White Oak Road to turn at Mr. Watson's gate and enter his fields to watch the racers raise the dust on a track just laid out. There had not been such a local concourse at Chatham since people gathered there to listen to the guns of First Bull Run.

Oliver Watson, Jr. sold Chatham farm in 1889 to Albert O. Mays of Ohio. It was already mortgaged. From the time of William Fitzhugh of Chatham onward the acreage had fluctuated but remained sizeable. From the shaky Watson and Mays ownerships the property went to others in parcels. When Fleming and Elizabeth Bailey of Georgia purchased Chatham in 1900, the house brought with it only a fraction over 30 acres. Chatham was no longer a large working farm.

The Baileys were succeeded in 1909 by A. Randolph and Frances R. Howard, of Fredericksburg, who lived at Chatham until 1914, carrying out improvements to house and grounds.

Chatham in the Early Twentieth Century

Next came Mark and Marie Sullivan. Invitation to use the Lacy pew in the Fredericksburg Presbyterian Church imparted a manorial stability to their Chatham tenure. Major Lacy had died in 1906 and his wife the following year. They retired to town during the 1890's, leaving Ellwood to tenant managers. Ellwood was sold in 1907. It is now a part of the Fredericksburg and Spotsylvania National Military Park.

Mark Sullivan, editor and author, one of the most distinguished journalists of the period, in the course of his career was associated with the *Ladies' Home Journal* and *McClure's Magazine*. He edited *Collier's Weekly*, 1914-1917, commuting from Chatham to his work in the North. In after years, he wrote the six-volume *Our Times: The United States, 1900-1925* and *The Education of an American*. Both the history and the autobiography are concerned

A Chatham Interior, Early Twentieth Century

with the relationship of events and ideas. In the latter, he traces his life from boyhood in Pennsylvania through Harvard schooling to a career of militant liberalism changing to conservatism.

In Sullivan's Chatham time, one of the Fitzhugh clan—the descendants of the first Virginia Fitzhugh were numerous—often visited Chatham, though not the house and grounds proper. He went to see a recluse who lived in a cottage across the big ravine north of the Chatham dwelling. He did not call alone. A gaggle of urchins and tomboys usually followed this retired Episcopalian rector in his traditional black garb topped by a pith helmet. His white beard hid his clerical collar but did not extend to the gravy stains on his waistcoat. A stalwart ex-Confederate, he would muster his little army at his cousin "Fish House" Fitzhugh's place of business near the Chatham Bridge to march off on what were called nature-study walks. Validating this designation, he occasionally pointed out obvious flora, wiping his face with a red bandanna and recommending the *National Geographic Magazine* for further study. Then he would tell funny stories about the war.

One day while he was deep in the slow talk of summer with his friend north of Chatham and his troops were scattered about reconnoitering, the young Sullivans appeared on the other side of the ravine and words developed between Town and Country. Stones followed. One of the Sullivans was wounded, and that party withdrew. Summoned to the scene by the cries of battle, Mr. Fitzhugh quieted his followers, bade them not exult and deprecated violence. Subsequently, with all the grace of a descendant of the First Families of Virginia, he mollified various headquarters, especially the one at Chatham.

Mark Sullivan wearied of commuting. In 1920, the Sullivans conveyed to Helen G. S. Devore, of Washington, wife of Colonel Daniel B. Devore, a Chatham advertised as having been thoroughly modernized by the current owners; it had its own water system, electric lights, hot-water furnace, and four bathrooms.

Helen Devore, née Stewart, wealthy daughter of a Wisconsin lumberman and Congressman, hired a Washington architect, Oliver H. Clarke, to create at Chatham an aura of colonial Virginia. The front porch walkways no longer existed. All of the front porches and also the back portico were now removed. At some time during the 19th century the elegance of the Flemish-bond brickwork with the glazed headers had been hidden by mustard-colored and later red paint. Though the original surface was still overlaid, white now softened the effect. Where once there had been wooden shingles, slate now covered the roofs. Carved limestone

Mark Sullivan

Daniel B. Devore as a Cadet.
Courtesy United States Military Academy

Garden Scene of the Devore Period.
Photograph by Miss Frances Benjamin Johnston. Courtesy of the Library of Congress

frontispieces, based on 18th-century models, now framed the main front and back doors, now opening on limestone steps.

Inside, the south stairway was removed—the north had already gone—and a large central staircase installed. Period woodwork was reconstructed where necessary.

On the grounds, tennis courts were eradicated and garden walls built. Specimen boxwood, box-bordered parterres amid carefully groomed grass, rose gardens, iris and other specialty plots, the whole set off by statuary, appeared in place of informal lawns and scattered shrubs. The demesne presented no longer the early 20th-century locale of tennis and croquet and hammocks where ladies might idle away long afternoons with the novels of Mary Johnston or Amélie Rives. The impression given by the Devore Chatham was English, a woman reporter wrote in a magazine article of the '20's. The green thumb here was indeed British. A professional gardener from Scotland, David Hanlon, who had worked on large estates in Britain and America, supervised the horticultural metamorphosis.

Mrs. Devore, herself a knowledgeable gardener, became the first president of the Rappahannock Valley Garden Club in 1924. She sponsored any interest in gardening and opened her premises to travel and garden writers. Pictures by Miss Frances Johnston, a noted architectural and environmental photographer, preserve the appearance of that period.

Colonel, later Brigadier General, Devore, West Pointer, veteran of the Spanish-American War and World War One, has a place in ecological history. A special assignment in the 1890's was to purchase reindeer for Alaska, a part of the introduction of that animal to the future state. Reindeer, the European and Asiatic domestic form of the caribou, became essential to Eskimo economy. At Chatham Colonel Devore raised pheasants and peacocks. He was remembered by Fredericksburg school children as a tall old gentleman courteously escorting their visiting classes around the mansion whose windows, gilded by the westering sun, shone across to their school directly opposite.

Chatham on its hill in the '20's seemed both far and close to the riverside citizens, who heard in those quieter days, instead of automotive sirens, the call of the Devore peacocks.

The Sullivans had added somewhat to their holdings, and the Devores acquired considerably more of the old Chatham acreage. In 1931 the Devores sold to John Lee Pratt, industrialist, an estate of some 256 acres, not counting numerous lots. The Devores moved back to Washington. General Devore died in 1956, at 95, the second oldest officer of the Army's retired list and the last surviving member of the West Point Class of 1885. Mrs. Devore died in 1960. They left no children. Her will provided for the care of sick and crippled children and research in the field of childhood diseases.

John Pratt made further improvements to the heating and water systems, but found no need for other changes. In decoration, whether inside or out, he deferred to his wife, the former Lillian Thomas of Philadelphia, whom he had married in 1918. She delighted in the Chatham gardens. After her death, 1947, her husband, though he too liked flowers and pleasant premises, maintained Chatham in simpler fashion and regretted the publicity of the '20's. He opened the grounds on occasion, but not the house. He discouraged travel writers. Local people, however, were invited to view the gardens on a spring Saturday afternoon each year.

Mr. Pratt had begun before the death of his wife a new career, that of philanthropy, in which Mrs. Pratt had joined. She was wealthy in her own right, a collector of antiques and art, and left to the Virginia Museum of Fine Arts at Richmond an outstanding collection of the last Russian Czar's family jewels, including examples of the precious Easter eggs by Peter Carl Fabergé. Such enchantments as a rock crystal egg rimmed with diamonds, a world globe of topaz on a gold base, and a bird of emerald in a golden cage found their way from St. Petersburg to Richmond by way of Chatham.

A widower without children, John Pratt, while not in his lifetime wishing his home to be considered a "shrine," as historic houses open to visitors were often called, did want Chatham to be the property of the nation after he left. He made arrangements for this disposal and, meantime, donated to the Fredericksburg and Spotsylvania National Military Park a half-mile strip of waterfront, 17.22 acres, including the upper pontoon site of Civil War occupation, part of the area of the battle of the crossing, December 11, 1862. When he died in 1975, the Park gained possession of the dwelling and 29.78 acres, since augmented by acquisitions bringing the total Chatham estate holdings to a fraction over 80 acres.

The Pratt executors sold much valuable furniture, silver, glassware, and art at auction on the premises, Friday, May 7, and Saturday, May 8, 1976. This sale, conducted by an international firm, drew a large crowd of collectors and dealers. Many local residents came only to watch, not availing themselves of the little ping-pong paddles that gave holders the right to bid. The paddles waved in good prices. Nothing went anywhere near as cheap as the lot of six pickle leaves for 15¢ at the auction of the late Churchill Jones' personal property in December of 1822.

Moving in, the National Park Service undertook an architectural study, both documentary and physical, and an extensive renovation for public use. The architects concluded not only that much of the exterior woodwork is original, but also that a good bit of the old interior remains, despite accounts indicating all but total depredation inside during the Civil War.

On October 15, 1977, Chatham and grounds were opened to offer an illumination of Colonial, Revolutionary, Early Republic, and Civil War History, with sidelights on the changing times since the Civil War. Exhibits indoors and out relate this story of an estate older than the Republic. The house itself, historic but young on its immemorial ridge, affirms life and continuity.

Rear View of Chatham

The Chatham Kitchen

Pratt of Passapatanzy

JOHN LEE PRATT never forgot Passapatanzy, his native place, an extended rural neighborhood centered eight or so miles east of Fredericksburg, Virginia. The music and mystery of the word claimed his allegiance. He knew also that some Indian names seem amusing to the Red Man's successors; so, up North, when asked where he came from, he did not say Fredericksburg, but Passapatanzy.

He was born at Aspen Grove in King George County, on a farm straddling Stafford and King George, October 22, 1879, a son of Alexander and Agnes Jones Pratt. Alec Pratt farmed Aspen Grove and also, with relatives, a large part of the south shore of Potomac Creek. As a member of the crack 9th Virginia Cavalry, young Alec had followed Robert E. Lee devotedly, remembering him in John's middle name. As a hard-pressed land owner in the Reconstructed South after the Civil War, Squire Pratt brought his children up to realize the value of work. There were four brothers, one of whom died young, and two sisters. The family lost the mother before John was six years old. (Some years later Alec remarried.) One of the brothers, Frank, later a distinguished surgeon, told an interviewer that the boys worked on the farm or at school year in and year out, with time off for sport fishing only twice a year: Easter and the Fourth of July. Presumably net fishing and crabbing sheltered under the aegis of work.

John Lee Pratt, following local schooling and a couple of years at Madison County's Locust Dale Academy, a private school now defunct, spent a year at Randolph-Macon College in Ashland, Virginia, proceeding then to the University at Charlottesville. He graduated, 1905, in civil engineering. During the course of his early professional years, he had taught for a time at a small Passapatanzy school. Decades later, if anyone mentioned an old native of the area,

John L. Pratt

the eyes of the multi-millionaire industrialist would twinkle and he would say, "I believe I taught him," sounding like a retired career school teacher.

The young engineer had the opportunity of joining a bridge company, but decided on advice of a professor to go with E. I. duPont de Nemours and Company. He worked with that historic outfit from 1905 to 1919 (except for the period of school teaching) and cherished the association his life long. In his old age, a chat with an acquaintance was interrupted by a telephone call that went on and on. Mr. Pratt kept smiling and making an occasional short remark. When he was finally able to hang up, he laughed and explained that the caller was an old duPont friend who liked to place expensive long distance calls when in his cups—"But that's all right; he's got plenty of money." One of Pratt's duPont assignments was a mission to Chile to expedite production of nitrate for explosives. Then in World War One he worked on various other special assignments.

After that war, duPont gave executive assistance to the General Motors Company. John Pratt transferred to G.M.C. The automobile industry, reaching adolescence, matured rapidly. With it rose Pratt. He became a director in 1923 and was vice-president from 1922 until retirement in 1937, though continuing to serve the company after that date. Squire Alec had possessed and been possessed by land, handling no inordinate amounts of ready money. John, one of America's richest men, could own land for his delectation. He held various acreages and liked to consider farming his occupation.

During the Second World War, a dollar-a-year man, he served the United States Government, visiting London for a time and savoring the grace under pressure of that beleaguered city. He did not fare so austerely or dangerously at Chatham, but he and Mrs. Pratt as a point of honor lived for the duration in an estate outbuilding to conserve energy.

Some of the many philanthropies of his widowerhood involved grants for the study of plant genetics. The ginkgo family, surviving since remote times, fascinated him. Whatever these inscrutable oriental trees withstood in ages past, they now seem immune to disease, insects and city fumes. Beautiful shade trees with fan-shaped leaves, they are desirable in the male, not the female, bearer of messy fruit. Chatham has a female. There is a male in the National Cemetery on the west side of Fredericksburg. Mr. Pratt, a great walker, used to stride down the hill from Chatham and on over the bridge and out through the town to the National Cemetery just to admire and assess the ginkgo there. He underwrote professional studies and gave a large specimen for experimentation to the University of Virginia. The ginkgo connection exemplifies a spirit of wonder. Mr. Pratt honored all life, all survivors, even the weeds he fought; he funded studies of them too.

Cultural as well as biological life concerned him. He helped individual scholars. He helped libraries and aided in the restoration of court and other records. He operated alone and through channels, notably the John Lee and Lillian Thomas Pratt Foundation. In his lifetime he gave land not only to the National Park Service, but to Stafford County and the City of Fredericksburg. The recreational area near Chatham, St. Clair Brooks Park, memorializes his friend Clair Brooks, stonemason, local historian and honorary mayor of Falmouth. Since the donor's death, his executors have turned over adjoining land for a local park named after Pratt himself.

Bequests in the Pratt will to schools and other institutions ran into the millions. On a simpler level, many people owed him personal gratitude. He did not think much of mere handouts, but would go out of his way to help a person find work and dignity.

After his own retirement, brother Frank, who then lived next to Chatham, occasionally joined John for the walks around town, the two-man progression stopping now and again to greet people. Mr. Pratt wore a distinctive walking hat, conservative in every way, but distinctive by maturity. A news agent near the bridge used to twit one of his customers about wearing a hat like John Pratt's. The paper seller missed the subtle difference that Mr. Pratt's had once graced a shelf of an exclusive New York haberdashery. The old farmer simply liked old hats, shaped to his comfort.

Sometimes he stopped at the Fredericksburg Court House and chatted with Mason Willis, clerk of the court, who had roomed next to him at Locust Dale Academy. They stood near the entry (no need to sit, just a short visit), and mapped the byways of boyhood. Sooner or later, Mr. Pratt would leave (perhaps because the clock stood at four, the adamant time of departure from that place), walking past the site of a meat market to which he had long ago sold quail and game for pocket money, and on over the bridge and up the hill to his evening meal.

Always as amiable to any acquaintance as to an old friend, John Pratt, proverbial plain old shoe (whether or not custom made), nevertheless kept his private life very private. George Washington's association with William Fitzhugh and Chatham is of record. Not generally known about is the hospitality extended by the Pratts to many 20th-century notables. Authors, industrialists, scientists, and statesmen such as George C. Marshall came and went without fanfare.

When he decided to give Chatham to the nation, Mr. Pratt feared disruption of privacy and serenity. He would live there until he died; only then would the deed take effect, but meanwhile details had to be worked out; the place could not simply be presented in years to come to an unprepared National Park Service. So, during 1963, before publicly donating the river acreage in 1964, he swore the agency to secrecy and began the paperwork for the larger donation. The secret was well kept; he suffered no invasion of privacy—except for the incursion of N.P.S. people, and he enjoyed that. All sorts of official planning rolled on. Local staff was in and out, and bigwigs from away came to inspect and confer. Some were allergic to poison ivy, which infested the woods beyond the grounds proper. Mr. Pratt delighted, it seemed to one group of planners, in leading them through patches of the stuff. The courtly host would never knowingly subject a guest to discomfort, but the old country boy smiled inwardly when he learned of the weakness.

As trips north and visits to the Virginia springs became too troublesome to be undertaken, as card-playing cronies died off, he remained spry and cheerful on his own turf. He seemed in his eighties a stalwart survivor of other times, like the ginkgo. He continued a modified social life, his correspondence and attention to philanthropies. He retained, too, a zest for the details of farm management. He liked to eat his own fowls and produce. People, old and young, interested him as ever; he entertained the small sons of his secretary with insect lore on strolls around the garden. And he loved to learn new things himself.

One such new thing was really an old report of a long-ago happening. Someone gave him a copy of a story in an August 1903 issue of a Fredericksburg paper: "Mr. and Mrs. Alex. Pratt, assisted by their friends, gave the usual old Virginia fish fry . . . words could hardly picture a scene more lovely for an outing than this ideal spot. The morning was spent in fishing, shooting and boating, while picturesque darkies stood over camp fires preparing delicious fish and crabs . . ." and so on. The throng of relations and friends, many of whom were named, had come to Whipsewasin Point on Potomac Creek from King George, surrounding counties, Fredericksburg, and Washington.

That, said Mr. Pratt, was one of the last of his father's fish fries, given as time went on more for the helpers and neighbors than for outside guests. Squire Pratt was getting old; his boys were beginning to fish in other waters. A bucolic period drifted to its close, a placid time, if not entirely as picturesque and satisfactory to everybody as the turn-of-the-century journalese implied.

Now, more than half a century later, himself the product of an era that was changing, John Pratt studied to benefit a larger group. He studied also in a personal sense for pleaure, which, he once said, gratified his physician, who believed that reading keeps the mind sharp. This engineer's inquiring mind rejected no branch of learning; scientific reports, the American Civil War, Virginia history, the rise of Chinese Communism, all were distilled in that alembic. Time, however, diluted the efficacy of therapy. Senescence blurred an extraordinary intelligence. There was no study toward the end, but the end came in peace and comfort. John Lee Pratt died December 20, 1975, two months into his ninety-seventh year. A private service was held at Chatham in the morning of Monday, December 22. His ashes lie in Fredericksburg's Oak Hill Cemetery. The legacies of his questing spirit continue to enlighten the living.

Fredericksburg and Steeples from Chatham, 1982

A Postscript concerning Ghosts

A VIRGINIA MANSION of note is supposed to harbor a ghost. Chatham has been granted a lady-in-white who first materialized in this century. A lonely periodic walker, she has not been seen for some time. Other ghosts, however, abound at all times. Though these spirits have not a body like man (or woman), a perceptive soul can sense their permeation of house and grounds.

Thus, a visiting boy or old man at the soldier's grave on the terrace overlooking the river may perceive more than the unobtrusive resting place of one unknown warrior.

A young mother looking at the sampler of the little Fitzhugh girls, so frail and vulnerable, may see through the faded stitching to a harsher world, even for the affluent, a chancy domestic environment where children (and women and men) died early.*

An epicure may imagine William Fitzhugh entertaining George Washington with cakes and claret in the glow of candles—what matter that sickness lay beyond the help of claret, that an open fire in a spacious room did not dispel the cold of a frozen night.

At the front of the house, a naturalist may touch the bark of the catalpa trees and consider the years of sunshine and rain that have nurtured and the storms that have buffeted these gnarled veterans, once saplings that escaped the blades of war. (A fellow veteran, a nearby hackberry, fell to a March wind of 1981.) One can reflect that a little Lacy or Sullivan or Howard boy, like so many boys before and since, might have tried to smoke the seed pods that give catalpa its common name, Indian cigar tree.

The mockingbird or the cardinal you may hear in the garden might have delighted a Pratt or Devore or Jones or Fitzhugh—not that particular songster, but essentially the same bird many generations back. His song is the song of his forebears, cheerfully enduring. Through him, as through many things, we grope toward the past, seeking ourselves.

*A facsimile of this sampler is on display at Chatham. The original is at Custis-Lee Mansion (Arlington House)

U. S. A.
UNKNOWN
1862

Principal Characters & All Owners

Fitzhugh Family & Friends

William Fitzhugh of Chatham

Ann (née Randolph) Fitzhugh (Nancy), his wife

Ann Randolph Fitzhugh (little Nancy), daughter, married Judge William Craik

Mary Lee Fitzhugh (Molly), daughter, married George Washington Parke Custis

William Henry Fitzhugh, son, inherited Ravensworth

Captain Benjamin Grymes of Eagle's Nest, nephew to William Fitzhugh of Chatham

Colonel, later General, George Washington of Mount Vernon, William Fitzhugh's long-time friend, absentee owner of Ferry Farm near Chatham

Dr., later General, Hugh Mercer, purchaser of Ferry Farm, killed in the Revolution

James Hunter of Pine Grove, near Chatham, owner of iron works

Outstanding Fitzhugh Horses: Kitty Fisher, Regulus, Volunteer, Tarquin, the True Whig, Chatham

The Churchill Jones Family

Brevet Major Churchill Jones, veteran of the Revolution, second owner of Chatham

Martha (Selden) Douglas Jones, his wife

Elizabeth (Macmurdo) Gibson, Martha Jones' niece

Patrick Gibson, Richmond merchant, Elizabeth's elderly husband

William Jones & Coalter Families

William Jones of Ellwood, Churchill Jones' brother, owned Chatham after Churchill's death

Hannah (Jones) Coalter, William's daughter

Judge John Coalter, Hannah's husband, purchased Chatham from William Jones

Janet Williamson, Hannah's afflicted daughter by a previous marriage

Elizabeth (Coalter) Bryan, Judge Coalter's daughter by a previous marriage

John Randolph Bryan, Elizabeth Coalter's husband

St. George Tucker Coalter, Elizabeth's brother

Lacy Family

Betty (Jones) Lacy, child of William Jones' old age, Hannah Coalter's half sister

James Horace Lacy, Betty's husband

Graham Gordon Lacy, one of their several children, left Chatham memoir, as did his mother

Some Civil War Figures

General Irvin McDowell, headquarters at Chatham, spring 1862, corps commander under George B. McClellan

General Rufus King, commanding a division under McDowell, advanced on Falmouth, spring 1862; headquarters Chatham, summer

John Gibbon, noted U. S. artillerist, promoted to infantry general; joshed at Chatham by Lincoln, 1862; headquarters there, 1863

Thomas B. Barton, ancient citizen, on committee meeting Federals, spring 1862, later one of several hostages

Colonel, later General, Herman Haupt, engineer, rebuilder of destroyed railroad line

General Marsena R. Patrick, military governor of Fredericksburg, spring and summer 1862; later Provost Marshal General, Army of the Potomac; often at Chatham

President Abraham Lincoln, visited Chatham, May 1862, reception for officers there

General Ambrose E. Burnside, headquarters at Chatham late summer 1862; later, as army commander, viewed pontoon laying from there, December 11, 1862

General Edwin V. Sumner, commanding Burnside's right wing, Battle of Fredericksburg, headquarters at Chatham, December 13, 1862

Clara Barton, later founder of American Red Cross, nursed wounded at Chatham, December 1862. No relation to Thomas Barton

Walt Whitman, American poet, visited Chatham hospital after Battle of Fredericksburg

Prominent Later Occupants

Mark and Marie Sullivan and children—Sullivan a distinguished editor and author

Helen (Stewart) Devore, wealthy renovator of Chatham and grounds

General Daniel B. Devore, her husband, veteran of Spanish-American War and World War One

John Lee Pratt, philanthropist, willed Chatham to nation

Lillian (Thomas) Pratt, his wife, discerning collector of art objects

All Chatham Private Owners & Dates of Ownership

William Fitzhugh (finished house 1771 - sold property 1806)
Churchill Jones (1806 - 1822)
William Jones (1822 - 1829)
John Coalter (1829 - 1838)
Hannah (Jones) Coalter (1838 - 1857)
J. Horace Lacy (1857 - 1871)
Oliver Watson (1871 - 1882)
Emma (Watson) Jones (1882 - 1883)
Oliver Watson, Jr. (1883 - 1889)
Albert O. Mays (1889 - 1900)
Fleming G. Bailey (1900 - 1909)
A. Randolph Howard (1909 - 1914)
Mark Sullivan (1914 - 1920)
Daniel B. Devore (1920 - 1931)
John Lee Pratt (1931 - 1975)

Acknowledgements

Some years ago, while assisting in the negotiations between the National Park Service and John L. Pratt, I made notes on the Lacy House from the *Official Records of the Union and Confederate Armies* and various Civil War reminiscences, but far from exhausted sources and did not pursue the earlier and later history of Chatham.

The experts on Chatham, whose work is implicit in this little book, are George Harrison Sanford King and A. Wilson Greene. George King, Virginia genealogist and historian, compiled extensive notes and copies of documents (colonial governmental records, court records, periodicals, letters, genealogical material) for his friend Mr. Pratt. Whatever George King did not treat, Will Greene did, as a park historian at Fredericksburg in the 1970's, filling out the Civil War and later periods. Greene cited the help of Lacy descendants Ellen T. Lacy of St. Joseph, Missouri, and Edith Boulware Carter (Mrs. Hanford A. Carter) of Fredericksburg in a published article: "Chatham, A Sense of the Past," *Northern Neck of Virginia Historical Magazine*, December 1979, vol. 29, no.1. I am further indebted to Agnes Gayle Harris (Mrs. Francis M. Harris, Jr.), a niece of Mr. Pratt's, for family data, and to her, Mrs. Carter, Messrs. King, Greene, and Mark Sullivan, Jr. for reviewing my manuscript, and my wife, Louise Williams Happel, for typing the first draft and emending the text.

Fredericksburg Park Superintendent James R. Zinck, Chief Historian Robert K. Krick and staff members Edmund J. Raus and David A. Lilley furnished valued assistance, including editorial review. Mrs. Joyce D. Brent typed the final copy.

Quotations from manuscript sources are by courtesy of the New York Public Library (Misc. Folder: William Fitzhugh), the Virginia Historical Society (Lee-Ludwell Papers and Grinnan Family Papers), the Virginia State Library (Gibson Family Papers, Accession 25342), Swem Library of the College of William and Mary (Brown, Coalter, Tucker Papers), University of Virginia Library (Bryan Family Papers #3400 and Fitzhugh Papers #5242—John L. Pratt donated the Fitzhugh Papers: letters of William Fitzhugh to Benjamin Grymes).

Statue of Ceres—Chatham